1-18-77

To James
With love
Mama and Dad

6-
5"

OLD FARM TRACTORS

OLD FARM TRACTORS

BY

PHILIP A. WRIGHT

WITH A COLOUR FRONTISPIECE
48 PHOTOGRAPHS AND 28 DRAWINGS

DAVID & CHARLES
NEWTON ABBOT LONDON
NORTH POMFRET (VT) VANCOUVER

ISBN 0 7153 5619 4

First published 1962
Second impression published by
David & Charles 1972
Reprinted 1974

Printed in Great Britain by
Redwood Burn Limited Trowbridge & Esher
for David & Charles (Holdings) Limited
South Devon House Newton Abbot Devon

Published in the United States of America
by David & Charles Inc North Pomfret
Vermont 05053 USA

Published in Canada
by Douglas David & Charles Limited
3645 McKechnie Drive West Vancouver BC

CONTENTS

ACKNOWLEDGEMENTS

THE author wishes to thank Mrs. F. R. Gulden for typing, and Mr. W. J. Priest for reading, the MS. Mr. Priest also kindly loaned catalogues, as did Mr. Norman Collings. Mr. J. G. Maddison supplied a Farm Tractor Handbook and Mr. Stanley Farmer of *Practical Power Farming* loaned valuable photographs from his files (and from those of the *Farmer and Stockbreeder*). Except where stated in the text, the other photographs are by Derek Evans, Messrs. Ransomes, Sims & Jefferies, the Ford Motor Company and Messrs. Perkins. All the other illustrations are from my own camera and sketchbook.

PHILIP A. WRIGHT.

Roxwell Vicarage,
Nr Chelmsford, Essex

ILLUSTRATIONS

The Plates

Drawings in the Text

INTRODUCTION

The art of ploughing has not altered a great deal since the prehistoric era. It is still possible to find ox and horse drawn ploughs today and occasionally the steam plough. Since the turn of the century, however, the internal-combustion engine has become the chief motive-power for ploughing and almost every other known farm activity. The tractor, in fact, was the inevitable outcome of the drift of labour from the countryside. Having already dealt in separate books with both *Old Farm Implements* and *Traction Engines*, I am satisfying a wish to place on record a note on the evolution and progress of the farm tractor.

After five years of experiment in 1902, Dan Albone produced the first really good farm tractor. The Ivel, in fact, continued to be made until 1921, and an excellent specimen is preserved in the Science Museum. Rivals soon followed; some were short-lived machines of great clumsiness, both of British and American manufacture. With the implications of the first world war and the entry of Henry Ford into the field, there was considerable expansion.

In 1957 the enterprising committee of the Suffolk Agricultural Show staged a parade of veteran tractors and the interest shown was tremendous. My photographs will prove that there are a large number of these ancients in preservation up and down the country, many being given the same lavish care as is bestowed on veteran and vintage cars.

I have felt it proper to trace the development in design up to 1930. After noting the implications of the 1914–18 war, with the U-boat menace threatening our food imports, we shall see the enormous influx of the motor plough, as we called those early tractors. We cannot ignore the American importations in this respect and I have tried to treat these separately. It seems fitting not to touch upon modern tractors at all.

It is not easy to recognize innovations in tractor design at the present day, and the difference between the earlier models and today's examples is a summary of hundreds of new ideas plus improvements of some very old ones. Tractors have improved greatly in design, flexibility and performance. The spade-lugged wheels have almost entirely been superseded by the giant pneumatic tyre and the caterpillar track. The latter was known very early in tractor history and the Holt caterpillars (upon which were mounted fighting-turrets) formed the basis of the very first tanks used in the 1914–18 war. The design of special implements for special tractors was also part of a later development and the hydraulic-lift system made control less exacting for the tractor-driver. The late Harry Ferguson was a pioneer in the latter respect. Fuel changes have seen the petrol-paraffin engine gradually giving place to diesel.

Pioneers have battled to give us the means of supporting millions of sheep, pigs, cattle, and poultry. Today the countryside is loud with the hum of machines all contributing to the feeding of the population. More than any other it is the farmer and his men who have made us what we are, and in this they have been helped by the engineering pioneers in every age.

I

THE PIONEERS

In considering the farm tractor the inclination is to begin with the self-moving type and to ignore the pioneering work which brought about a greater appreciation of the internal-combustion engine in the field. I feel, however, that this would not be right, in view of the success of the early oil engines, which were not only bedded down in barns to drive farm machinery but were made portable by mounting on wheels to be drawn to their place of work by horses, evolving in the same way as the steam portable engine. In 1890, Herbert Stuart, a Yorkshire engineer, with Charles Binney, produced a highly successful oil engine, designed to run on paraffin or ordinary lamp oil. The engine had neither slide valve, igniting tube, nor electricity, but a blowlamp was needed to start the engine from cold, after which it would run for an unlimited period with the lamp removed. A year later Richard Hornsby & Sons of Grantham took up the manufacture, and their Hornsby-Akroyd patent oil engine was on sale before Dr. Diesel had produced anything.

The pioneer oil engine was a success and by 1897 was being sold everywhere. Three years before this the Royal Agricultural Society held a series of oil-engine trials at Cambridge, and Hornsbys led the field.

Following the appointment of Mr. David Roberts as managing director, the firm decided to stop their steam-engine building, as the internal-combustion engine appeared to offer tremendous prospects. Although these engines continued to be built for a long time, it was generally realized that the farm power of the future must be self-moving, and it was left to others to produce the prototypes of the motor plough and farm tractor.

Dan Albone began life as a true son of the soil, his father being a market-gardener at Biggleswade, Bedfordshire. Dan's zest for things mechanical soon took him into other realms, however, and after a spell of cycle racing, he took up the repairing and manufacturing of cycles—rather in the same way as the early start of the present Lord Nuffield. When still only twenty, Dan issued his first catalogue, and by this time he was supplying ball-bearing hubs and wheels for horse-drawn traps and carriages. This in itself was an innovation. With his roots still in the soil, Dan had the foresight to see that the motor had a future on the land, where horses and steam were at that time the only forces.

In 1897 he began experimenting, and by 1902 the first Ivel tractor was being tried out. There were some distinguished supporters, including S. F. Edge and Charles Jarrott, who became directors of Ivel Agricultural Motors Ltd., formed in 1903 with a London office and a factory at Biggleswade. Dan was constantly improving the tractor and found time also to make a potato-planter and a prototype motor fire engine. By 1906 he was demonstrating harvesting at night, when the Ivel drew two binders and cut seventeen acres in six hours. That same autumn Dan died and in 1921 his company succumbed in a period of economic depression. In all, 900 tractors had been made, a large number of them being sent abroad; the British tractor industry was launched.

The Ivel was a twin-engined tractor of $4\frac{7}{16}$ in. and 5 in. stroke, which could develop 24 h.p. The surface carburettor drew air by engine suction through a fuel reservoir, then inlet and outlet valves were spring-loaded and automatic. Coil ignition was fitted and a cooling water circulation system embodied a 30-gallon tank, but there was no radiator. The machine even had a power take-off pulley and was certainly both imaginative and successful. Control was by a hand lever for ignition, with clutch and brake. When the plough was lifted, the tractor accelerated and shook extra fuel on to the wicks of the carburettor, and the air-valve throttle would need to be hurriedly closed. One control lever was used for forward, neutral and reverse, employing a massive cone clutch. Drive to the rear wheels was by multiple chains and sprockets, the rear axle being fitted with a differential. A beautifully preserved specimen is in the Science Museum, and Mr. W. O'Dea, the Keeper of the Agricultural Department, who supplied the photo, tells me it lay unnoticed for many years with a

The 1902 Ivel tractor.

Ransome's self-propelling oil engine.

Sharp's 1904 unique model.

Ransome's 1902 tractor.

lot of rubbish. Like many others, the tractor is very light at the front, and there was an alarming tendency to overturn on a big pulling job. Nevertheless, this was a remarkable effort and Dan Albone was nothing short of a genius. At the 1904 Royal Show his tractor was awarded a Silver Medal.

In 1903, Ransomes, Sims & Jefferies Ltd., of Ipswich, built an agricultural petrol motor-tractor. The firm was already world-famous for farm implements and steam engines. This tractor was driven by a 20 h.p. four-cylinder engine and was not unlike the cars of that period in appearance. There was a friction clutch with chain drive to the rear wheels, and there were three forward speeds and reverse; two brakes were provided, one a foot brake and the other a hand brake acting upon the road wheels. An extension shaft through the gearbox carried a pulley for stationary work.

It was claimed that the tractor would plough, cultivate, and tow binders, reapers, and mowers, in addition to driving threshing machinery, pumps, and saws. It was also claimed that this machine would, with a three-furrow plough, do five acres in ten hours, and draw seven tons on a level road, in low gear. Quite soon, however, the company declared that the support received was insufficient to encourage manufacture and so the projected £450 tractor was dropped—perhaps the farmers of that day thought it a bit expensive! About ten years later Ransomes offered an oil tractor for farm work and general haulage. This machine weighed some ten tons and was rated at 35 to 40 h.p. There was only one cylinder, but this was 12½ in. diameter by 14 in. stroke; overall length was 18½ ft., width 7 ft., and height to top of exhaust chimney 11 ft. She could run on either paraffin or crude oil, but her massive size precluded success for direct field traction, of course.

In 1904 Sharps Patent Auto Mower and Tractor Company, operating in York, produced an amazing little tractor, one of which will be seen in a later chapter to be still at work. It was originally powered by a twin-cylinder Daimler engine, which was later replaced by a four-cylinder Humber. This very early tractor actually incorporated a power-driven mid-mounted cutter bar for grass mowing. Through the kindness of Mr. Stanley Farmer of *Practical Power Farming*, I am able to reproduce a contemporary photograph and also quote from a maker's advertisement which claims that the tractor "will mow 35 acres of grass in 10 hours. Cost of fuel—2d. per acre. Will haul 3-furrow plough at 5 m.p.h. Will haul reaper, binder, and other farm implements. Will drive threshing machine, chaff cutter, bone grinder, cream separators, churns, laundry machinery, or anything not requiring more than 14 h.p. Will haul 2 tons over any field, and 5 tons along a road." This little tractor weighed fifteen cwt.

In 1907, Blackstone & Co. produced an "oil traction engine" with a four-cylinder engine. *Farm Implement and Machinery Review* said at the time: "For motors of this description, for agricultural purposes, a great future is known to exist." There was, however, little encouragement from the farmer.

Marshall, Sons & Co. Ltd., of Gainsborough, Lincolnshire, made an early agricultural oil motor at this time, but it weighed some four and a half tons. The two cylinders were cast in one piece with a water-jacket and the horsepower rating was 30. It had the appearance of a steam engine and was equipped with a driving-pulley. At the Royal Show at Gloucester in 1909 among the new implements (awarded a Silver Medal, too) was the Cyclone tractor, made by a London firm, Cyclone Agricultural Tractor Company. Its special feature was an arrangement for driving a mower direct from the engine shaft by means of an eccentric. Fitted with a 20 h.p. Aster engine, she could plough and also drive by pulley any stationary farm machinery.

In 1910 the Royal Agricultural Society of England organized the first competitive tractor trials in this country. These took place during six August days at Bygrave, Baldock, Hertfordshire. There should have been eleven entries, but only seven arrived, and three of these were steam tractors. Of the four, two were Ivels and two were by Saunderson & Mills. Both Ivel tractors were 18–20 h.p.—one a single speed and the other a two speed. The Saunderson & Mills exhibits were of their Universal model, one being rated at 25–30 h.p. and the other one 45–50 h.p. The larger model had a four-cylinder engine and the fuel was paraffin. I mention the number of cylinders, because, for a long period, they varied from one to six on early farm tractors. The drive was taken through a cone clutch to the gearbox and thence via spur wheels. The belt

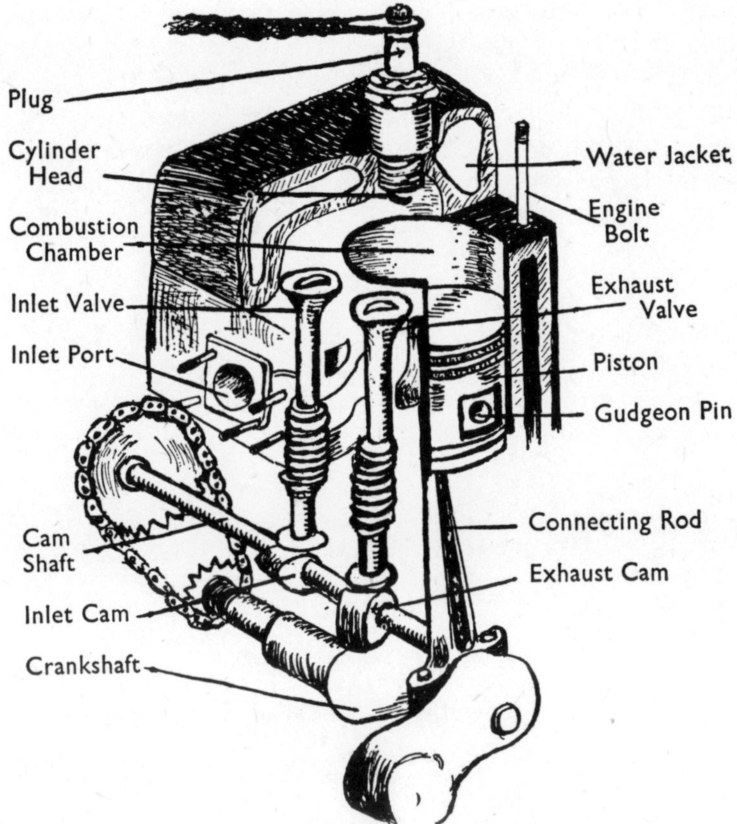

Fig. 1. Typical four-stroke layout

pulley was on the end of the clutch shaft. The tractor was priced at £450. Mr. Howard, the present managing director of Howard & Dennis Ltd., Bedford, was chief engineer of Saunderson & Mills Ltd. for a long time, and he has recently told Mr. Jim Priest of *Farm Implement and Machinery Review* that it was not merely the initial cost of these tractors which frightened off prospective customers. There was a strong fear among farmers of having petrol about the place and, of course, petrol was needed to start the tractor and warm the engine before switching over to paraffin.

The R.A.S.E. judges took some months to produce their report, and when they did so they came down heavily on the side of the steam-engine entrants. They said that they were "unable to specify the ideal agricultural motor". They wanted to see one "which, while not being heavy, would enable the farmer to dispense with the hiring of a traction engine and would be capable of doing a substantial amount of field work". We shall see in the ensuing chapters that such tractors were indeed produced.

II

THE IMPACT OF WORLD WAR ONE

IT is often supposed that until the first world war these tractor and motor-plough pioneers had marked time. Although the war gave them a tremendous fillip and helped to dispel the distrust felt by conservative-minded farmers, quite a lot of good machines were produced after the 1910 trials. An early catalogue of Saundersons shows how they continued to improve their Universal tractors, and certified tests were made on Mr. Joseph Newman's farm on October 13th, 1911, at Harrowden, Bedfordshire, and the following February on the Earl of Radnor's Yew Tree Farm, Salisbury, where in twenty-four hours over twenty acres were ploughed continuously during a period involving a wet night. In a contemporary photograph I am glad to notice a parson among the spectators.

In August 1914 the Universal range was tried out for harvesting operations at Village Farm, Stagsden, Bedford. Model "J" hauled one binder easily and two satisfactorily. Model "G" hauled two binders easily, model "B" hauled three, and model "S" hauled five binders, and there are excellent photographs to prove it. Saundersons were also exporting a great number of tractors, and these Elstow-made tractors were by then in Russia, New Zealand, Italy, and Africa. A wonderful haulage run in March 1914 took place from Londiani to the Uasin Gishu Plateau, which involved a 2,000 ft. climb. The journey of sixty-three miles was completed in twenty-one and a half hours on twenty-eight gallons of paraffin.

Nearer home the Jockey Club at Newmarket was successfully using a Universal model "G" for towing three 6-ft.-cut grass-mowers and then a 10-ft. roller weighing three tons, plus a set of harrows, on the racecourse. The club began using it in 1911 and two years later was full of praise for its performance.

The outbreak of war found the farmers suddenly thrust into the forefront after a long period of indifference on the part of the general public and successive governments. The plain facts, of course, were that the country was stunned and frightened. The need for greater production was intensified by the successful work of German U-boats in checking our imports. In this tremendous demand for home-grown food, the importance of tractors cannot be underestimated. The great increase of acreage which it was possible for one man to till and harvest was almost incredible. Take for instance an eight-acre field of our stiff Suffolk clay—one man with his horse team might take eight to ten days to complete the ploughing. With tractors and ploughs, even of this 1914 vintage, a man could plough the whole field in one day. The tractor output was even increased in some cases and one and a half acre ploughing per hour was not unknown. Most of the tractors and motor ploughs had a seat for the driver by now and so he became less tired than when following a horse team.

The really big advantage of the tractor power went even further, and meant briefly the ability to do the maximum amount of work at the right time. Ploughing at the right time and under the right conditions could, of course, reduce labour for the after-cultivation, and even drilling by tractor power was coming in at this time. Stiff lands were ploughed very deep, something which only the steam plough had achieved until this time. Rolling, harrowing, reaping, and cutting by binder were all practical propositions.

There was, of course, opposition, and well into the 1920s I heard farmers and their men protest that the tractor wheels packed and panned the land, whereas the horse trod lightly. There was a good deal in the criticism, especially with the use of some types of wheel, and, as will be seen later, considerable ingenuity was brought to bear upon improving wheels. In May 1917 an Ivel tractor was demonstrated on a farm owned by the celebrated engineer, Mr. S. F. Edge, in Sussex. This 20 h.p. tractor came on to a field of stiff Weald clay which had already been ploughed and had set into solid lumps. In a very short time a good seed-bed was produced. There still are times, however, when the tractor brought out too soon on wet land does more harm than good. Even in the wet year of 1960, the few remaining farmers who had horses and horse implements got their corn drilled earlier by horse power. Farmworkers were also being called up for military service in 1914 and horses were commandeered. The need to save labour was therefore another important factor in favour of motor farming as it was called.

Under Government orders, the Ministry of Munitions built an M.O.M. tractor. It was not intended to market this machine, but to hire it out to farmers. It seems to have been a serviceable tractor and stood its test well, but it was never developed. Instead, in 1917 Lloyd George (then Prime Minister) asked the late Henry Ford to ship 7,000 Fordson tractors to this country. I shall deal with this in a chapter devoted to American influence. The Government, however, did own several hundred tractors driven by soldiers and land-girls. The first of these to come to Bryers Farm, Hawstead, where I was a youngster, was very soon ditched, and it must be admitted that many of the drivers were inexperienced.

A variety of makes of tractor were therefore in use in the early days of the war. We called them "motor ploughs" at first—some were little more than a motorized plough—i.e. a mechanical horse. The number of cylinders varied from one to six, paraffin was the usual fuel, and the engine would be started on petrol until warmed up. The little self-contained motor ploughs turned in a very small space and worked best on tiny irregular fields. There was also a fine sense of balance, and as in horse implements, the operator sat behind and could see his implements in front of him. The weight displacement, too, was borne mostly by the driving-wheels, thus ensuring a good tractive grip on the ground.

The Highland Agricultural Society of Scotland was able to stage war-time trials in 1917 and I would like to quote the judges' report dealing with weight. "With regard to weight it will at once be apparent that in many cases the tractors shown have been evolved on the lines of the heavy steam tractor. The most noteworthy feature of the present demonstration was the appearance of the light land tractor constructed on new lines. Several manufacturers appeared to have departed from the idea that great weight is necessary for a tractor to do efficient work on the land. It was clearly shown at the demonstration that light machines, adequately provided with speeds, grip the ground and perform the work better than the heavier machines. Every drawback, such as slipping on soft land and inability to climb gradients, was aggravated by increased weight above a certain limit. Extra weight also increases the risk of breakage where stones are encountered and the danger of injury to the land through compression, this being very noticeable during the demonstration, especially when the driving-wheel on the land travelled within a few inches of the previous furrow. Besides this, a heavy tractor is at a distinct disadvantage for the other and lighter forms of cultivation, such as grubbing, cultivating, seeding, and harrowing, and also for harvesting, especially where the land is sown with grass seeds.

"The light tractor is quite suitable not only for ploughing, but for other farming operations, and therefore embraces all the usual requirements of a farm tractor including the driving of a threshing mill and other farm machinery. It should be kept in view that for heavy stationary work proper anchorage must be provided. The only class of work for which the light tractor does not appear to be suited is road haulage. For road haulage, weight is a necessity, as high speeds cannot be used to give the necessary gripping power. This leads the Committee to the conclusion that field work and road work are more or less incompatible, and that a light tractor, such as is suitable for land work, is not so useful for road haulage. In addition, for road work a tractor should be sprung to minimize the vibration, while for field work, springs are undesirable, and only entail extra weight. The Committee have come to the conclusion that, to suit conditions in Scotland, an efficient land tractor need not exceed 30 cwt. in weight. In point of fact, several of the tractors which did excellent work at the demonstration were well within that weight."

In the spring of 1918 the Government offered, through the Food Production Department, a Championship Shield for the best acreage ploughed by any tractor in three consecutive months. There were 1,700 competitors and almost every tractor then in use in this country entered the contest. The two-speed Overtime model N tractor was the winner, and it was claimed that it could be easily handled by a boy or a land-girl.

The advantage of a separate tractor and plough was that quick disconnexion enabled the use of all types of implement for all-the-year-round farm work, a pulley power shaft for threshing, wood-

A 1917 Saunderson still in use.

A 1920 Saunderson still used.

The early Austin.

Another Austin at work.

The "General Ordnance" tractor.

A 1916 Avery in poor shape.

PLATE 6

A Killen-Straite 1917 model.

1919 Clayton track-layer at Lincoln Trials.

sawing or grinding and the ability to do cartage on road or farm. Two forward speeds were a great advantage, a very low gear was of inestimable value for abnormally difficult work. Designers had a big problem in attempting to provide the greatest possible power within certain limits as to maximum weight. If fuel consumption were equal (plus quality of work), then the best tractor was the one which compressed the soil least of all. The self-laying track of the Bullock & Holt—of the kind known today as "Caterpillar"[1] types—had particular claims in avoiding panning the land.

In my book *Traction Engines*, I have described the Hornsby steam caterpillar, built for the Yukon. There is always a danger in producing anything in advance of its time and in 1908 Hornsbys had actually made a similar tractor powered by a twin oil engine. This would cross even a river with its chain-track efficiency. Hornsbys made preparation in Lincoln for a spate of orders which never came. So small was the interest on the part of agriculture that this venture became a financial disaster. Hornsbys therefore sold the patent rights to the Holt Caterpillar Tractor Company of New York, who persisted with its development. In January 1915 the First Lord of the Admiralty, Winston Churchill, wrote to Mr. Asquith, Prime Minister, suggesting small armoured shelters holding men and machine-guns to be mounted upon Caterpillar tracks, by which system trenches could be easily crossed and the weight of the machine would crush wire-entanglements. A year previously, Major-General E. D. Swinton had made a similar suggestion, and the traction was based on the Holt Caterpillar. So the rejected Hornsby idea was imported from Holts, to whom Hornsbys had sold their patents in 1908. The first "tanks", as they became known, were tried out on the Suffolk Breckland near to Brandon and Thetford.

There was great variety in the wheels of these early tractors. With the four-wheeled types, some ran one or more wheels in the furrow and some ran all the wheels on the land. Of the three-wheeled types, some were like the motor-cycle combination, and others had two steering-wheels in front with a large driving-wheel at the rear, or merely a front steering-wheel and two rear driving-

[1] The trade name of Caterpillar Co. Ltd. Generally, track-laying tractors are now known as "crawlers".

wheels. Many of the early models required a man to ride on the plough, cultivator, mower or binder. The self-lifting mechanism controlled from the driver's seat was therefore a welcome innovation and saved a man's time and labour, and indeed became an economic necessity. I will record some of the makes of this war-time period, but shall deliberately leave out those which appeared at the 1919 demonstration which is the subject of my next chapter. Some makes were very short-lived, like the Broom & Wade effort of 1914. Some agents without any agricultural knowledge were attempting to push business and by poor demonstrations did more harm than good; their work disgusted many farmers.

The Bates Steel-Mule was a modification of a three-wheeler, but instead of a big driving-wheel, an endless track was substituted. She was a cumbersome four-cylinder machine, as is seen in the photograph. Transmission was by sprocket and chain, she weighed two and a half tons and was rated at 30 h.p.

The Bullock-Creeper was a self-laying track machine from the U.S.A. which could turn in its own length. Its four-cylinder engine developed 35 h.p. and there was only one forward speed, but the weight was nearly four tons. The firm of John Wallace & Son, of Glasgow, produced an interesting machine called the Clydesdale.

Another British-built four-wheeled tractor was the Ideal, which had an almost modern appearance. It had self-drive mechanism for operating reaper and binder with a special mower attachment. The engine was four-cylinder vertical 35 h.p., and transmission was by toothed gearing and chain drive to the rear wheels. Its weight with plough was four and a half tons, and this may be the reason why little was heard of this machine after 1919. The splendid Ivel (then Ivel-Hart) noted in my first chapter also seemed on the decline, but had done wonderful work, and I have a photograph dated 1910 showing one of these drawing a binder. The International Harvester Company were also sending into Britain their Mogul, a 16 h.p. four-wheel model with optional extension rims for bad working conditions. This was a single-cylinder 16 h.p. chain-driven tractor weighing two and a quarter tons, and many of these were hired out by the Government. Another American import was the 16 h.p. Rumely, a three-wheel tractor with two wheels in front, the near-side one being an idler wheel. Steered from the rear, this looked an attractive project. The driving-wheel was hollow-straked and very wide, while a power lift could raise or lower the plough draught. She was a vertical four-cylinder engine and toothed gearing was enclosed and ran in oil; her weight with plough was about three tons.

A very little-known tractor should be recorded here. I refer to the Scott made by J. Scott of 12 North Saint Andrew Street, Edinburgh. It was an enterprising combination of tractor, cultivator, and seed-sower, weighing only twenty-five cwt. and costing £300. Mr. J. G. Maddison of Bury St. Edmunds tells me it appeared at the Suffolk show at Chediston Park, Halesworth.

In *Farm Implement and Machinery Review*, a Mr. William Taylor of Linnyshaw Moss Farm, Walkden, Manchester, reported that he had made a tractor and wished to negotiate with a manufacturer for its production, but I am unable to discover anything further about the project.

The British-built Wyles motor plough was very like the Crawley Agrimotor which I shall describe in the next chapter. The Wyles was a single-cylinder 11 h.p. type, and the ploughman walked behind and held the plough handles as though it were horse drawn. Built in Manchester, this grand little tractor weighed a ton. At this time, the Vulcan Car Agency, of Great Portland Street, London, were importing the Denning tractor, which had extremely wide extension wheels for very soft ground. The Austin Company, of Birmingham, produced a Culti-Tractor which was a three-wheeler with a one-wheel drive. Rated at 20 h.p. with a four-cylinder vertical engine, she had toothed gearing and weighed thirty-five cwt.

The American Goliath made no headway in this country, as she was too heavy. She weighed nine and a half tons and had a four-cylinder four-stroke 70 h.p. engine, toothed gearing, two forward speeds and a belt pulley for stationary work.

By the end of the war most farmers had decided that tractors had come to stay. The realization was made with great reluctance, as farmers often like to be left alone. Such an attitude can be understood up to a point and by then there was a bewildering number of machines on the British market. Advertisements and demonstrations of single tractors were common enough, but apart from the initial cost, a wrong choice could easily spoil a whole year's work on a farm. The Society of Motor Manufacturers and Traders therefore decided to stage some worthwhile trials and invited the National Farmers Union to appoint the judges. The trials took place at South Carlton, Lincolnshire, from September 24th to 27th, 1919, and my next chapter will give a brief report and a mention of every tractor there, whether British-made or otherwise.

III

THE TRACTOR TRIALS OF 1919

THE organizers of the trials at South Carlton, Lincoln, really set out to help the farmer to make a choice, and they held a firm conviction that there was a tractor to suit every farm. They therefore tried to get every tractor of importance represented, even if made abroad, providing it was actually on the British market. Ploughing, cultivating, threshing, and hauling were included as tests, but naturally the purely motorized plough did not participate in belt work. In addition to the farmer judges (six in all) a consulting engineer was engaged to make exhaustive examination of the tractor's structural features. No prizes or medals were awarded as in the case of R.A.S.E. trials, as the organizers felt there was no such thing as a "best" tractor, nor would there be in the foreseeable future. Tractor entrants were ordered to provide their own ploughs and cultivators, and steam-driven tractors were not excluded.

I think it best to list the entries in alphabetical order with a brief note of their qualities:

Messrs. Alldays & Onions, of Birmingham, had three similar models of their 30 h.p. Mark II tractor. She had four cylinders and the final drive was by chain. This four-wheel tractor had a gross weight of three tons, and was priced at £630.

The Austin Company of Birmingham had by this time produced a four-wheel model of 25 h.p. with four cylinders, spur-wheel drive, weighing one ton eight cwt. laden. At £300 this was outstanding value. An uncle of mine (the late Arthur Honeywood) purchased one during the 1914–18 war, and my cousins were using it at Norton, Suffolk, until a very short time ago. There were two of these working at the trials, as well as a 30 h.p. model which demonstrated threshing and hauling. In the main, the Austin bore a really striking resemblance to many modern machines, and at these trials it was described as easy to handle and safe in operation.

The 28 h.p. Avery had four cylinders, spur-geared transmission and a belt pulley. She weighed three tons five cwt. and sold at £500. She was entered by the firm of R. A. Lister & Co., Dursley, Gloucester.

The Blackstone (built at Stamford, Lincs) was rated at 25 h.p., and two models were at work, one on wheels and the other on tracks. These models had only three cylinders, weighed two and a quarter tons and sold at £500.

The 35 h.p. Clayton was built by the famous traction-engine firm of Clayton & Shuttleworth, Lincoln, had four cylinders and was governed. She sold at £650.

The famous U.S.A. firm, J. I. Case of Wisconsin, had an 18 h.p. and a 27 h.p. model, each with four cylinders and weighing one ton thirteen cwt. Neither model was spring mounted and the prices were £375 for the smaller tractor and £475 for the larger model. The Cleveland Company, H. G. Burford & Co. of Regent Street, S.W.1, had three models present. Each had four cylinders, weighed one ton eight cwt. and sold at £397. The horsepower rating was 21.2. This tractor named Cletrac was one of the smallest and most compact track-laying models of its generation. The engine was water cooled, mounted on to a main frame. The sub-frame carried the track wheels, and was linked to the main frame by a laminated transverse spring. A most peculiar feature of this little low-built machine was the belt pulley, which extended in front of the radiator. Drive being at right-angles to the tractor line made setting up for stationary work a very tricky business.

A fine little Essex-built tractor was the Crawley Agrimotor, built by the two Crawley brothers at Saffron Walden. It had already earned a great reputation for efficiency. There were two driving-wheels, one of which ran in the furrow; the driver sat behind and steered, and other farm implements or haulage wagons could easily be fitted. She was a four-cylinder 30 h.p. tractor and had two forward speeds; transmission was by toothed gearing, and she weighed thirty-eight cwt. and sold at £500 complete with plough. Mr. S. W. Crawley has recently written me from Harwich and tells me he built an experimental model of this excellent tractor at the age of fourteen, just after leaving school. A considerable number of the machines were exported between 1913 and 1924 as far afield as Australia. They were powered by a Buda or Puterber engine and would run on petrol or paraffin. Mr. Crawley's brother farms at Hadstock, near Saffron Walden, and both these clever engineers are most humble about their grand little Agrimotor, and have pointed out that they did design an experimental prototype as long ago as 1908 which was not proceeded with and which Messrs. Hedley & Edwards of Cambridge, built for them. Mr. F. A. Standen, of St. Ives, exhibited one of these at the Huntingdon Show in 1913. The ploughmen of those days referred to the controls as the "reins", as they were very much "horse"-power-minded at that period. This business of having the power and the work in *front* of the driver was excellent, and I personally have often wondered why such a system was not used in the modern tractor. The ploughman who does not look back (St. Luke 9, verse 62) is at least able to see where he is going without constantly getting a pain in the neck!

Another interesting tractor in the Lincoln trials was the Emerson-Brantingham, entered by Messrs. Melchior, Armstrong & Dessau, of Great Marlborough Street, London. It had four cylinders, the final drive being by enclosed spur wheels. A belt pulley was fitted, she weighed two tons and sold at £446. George Garrard of Gislingham, Suffolk, bought one in 1927 and used it with a three-furrow plough.

The Eros tractor unit was an attachment whereby a Ford car model T could be converted into a tractor for £65. Amazing claims were made for the strength and adaptability of this unit, as a two-furrow plough, binder, cultivator, harrow, and roller. Fitted with a pulley it would drive a circular saw. When not required for tractor work, the Ford could be made roadworthy as a car in twenty minutes. By fitting a Russell vaporizer it could even be run on paraffin. Messrs. Morris Russell & Co., of Great Portland Street, London, marketed this device.

The firm of Fiat Motors Ltd., 5, Albemarle Street, London, had two of their 25 h.p. Fiat tractors; one ploughed and cultivated whilst the other did some hauling and threshing. A four-cylinder tractor with final drive worm and wheel, it was a light tractor (one ton six cwt.) with four wheels, but no selling price was given. It was made in Turin, Italy.

The Fordson will be mentioned again when we review American influence, and it will be seen from these photographs by comparison that the general appearance of this world-famous tractor today is very little altered from the first model. Henry Ford had, in fact, been experimenting with a series of prototype tractors culminating in Model F, which although ready in 1913, did not come to the world's notice until four years later. Rated at 22 h.p., the Fordson had four cylinders and worm and worm-wheel drive. It was light in weight (one ton three and a half cwt.), almost too light in the early models, for there was a strong tendency for the front wheels to lift during heavy pulling. From the very first the Fordson was an attractive proposition at the comparatively low price of £280. Of the three Fordsons present, one had a belt pulley (£12. 5*s.* extra) and did threshing. By this time Ford had a factory in Cork, Ireland, and these tractors were made there.

A neat-looking little tractor was the Garner made in Birmingham and rated at 27 h.p., weighing one ton fourteen cwt., with transmission of worm and wheel. Chain drive was on the way out, as can be noticed. The price was £385, and a second model was there for threshing.

The G.O. tractor was made in Birmingham and marketed by Stockwells of Norfolk Street, Strand, London. There were four cylinders with final drive spur gears to live axle; one and a half tons in weight and usual four-wheel layout. She was nicely enclosed from weather and dust, and the selling price was £480, which, we are told, "included £30 worth of spare parts".

From the D.L. Motor Company of Motherwell came a Scottish-built tractor, the Glasgow, rated at 27 h.p. A four-cylinder engine transmitted power to the rear wheels by bevel pinions and spur wheel. Weighing one ton sixteen cwt., this tractor sold at £450.

The American Tractor Company of Avenue de Bel Air, Paris, had a very powerful machine, the Gray, of some 36 h.p. There were four cylinders and the final drive was by roller chain. A belt pulley was fitted, but the forbidding weight was no less than two

Bates' "Steel-Mule".

Clydesdale and Moline tractors on trial.

Fowler's cable oil-engine.

McLaren's windlass oil-engine.

tons fifteen cwt., and, of course, the price was high—no less than £600 being asked.

Only £5 less was being asked for the American Illinois tractor, which had four cylinders and spur-gearing transmission. Her laden weight was two tons eight cwt. In the preceding chapter I mentioned the Mogul 30 h.p. sent over by the International Harvester Company, who then had a London office in Finsbury Pavement, E.C.2. There were two of these tractors entered, still only twin cylinders, priced at £580.

The same company by this time had produced what was to become a very popular tractor in the Eastern Counties. Known as the International Junior, and rated at 22 h.p., it was a good proposition for the smaller farmer. There were four cylinders and with a weight of one ton sixteen cwt. she was easily handled. The price was £300 and the second model present at the trials did some hauling and threshing.

The same intrepid firm also produced the 25 h.p. Titan and, in fact, sent 3,500 Titan tractors into Britain from 1914 to 1920, when the model was discontinued. It was a twin-cylinder tractor with two forward speeds and tooth-geared transmission which ran in an oil bath. The weight was two and three quarter tons and

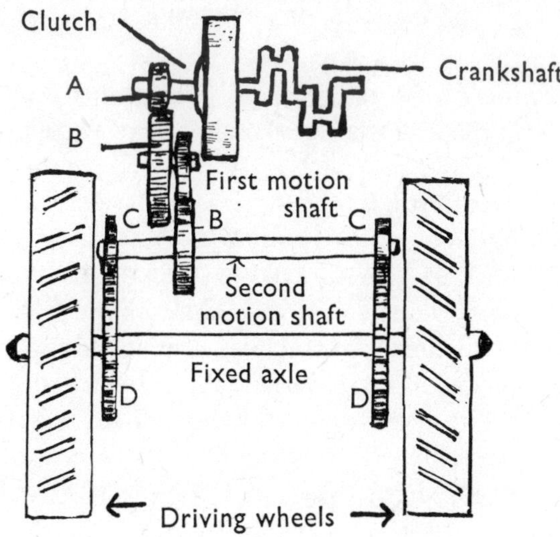

Fig. 2. Transmission system

the selling price £410. There are a few of these popular tractors still at work, and in 1956 a splendid model was given to the Museum of English Rural Life at Reading. There were two others of the same make at these 1919 trials. More than any tractor, the appearance of the Titan was very like a small traction engine minus its chimney. In place of a smoke-box, there was a 32-gallon rounded water tank. The tractor did not have a radiator, water was instead circulated through a pipe from the bottom of the water tank, round the cylinders and back via an overhead return pipe. The engine had a half-compression device for easy starting.

Entry number 36 was a Mann Steam Cart and most properly belongs to my *Traction Engines* book. It was designed, however, as a steam tractor for direct traction, rather like Garrett's Suffolk Punch. It was a smart little compound steam engine priced at £825. One of these steam tractors, owned by Mr. R. B. Haigh of Thaxted, has been beautifully reconditioned by Mr. A. L. Frost of Takeley, Essex, and his son, Peter. She regularly visits rallies in the Eastern Counties and was built in 1921 at Hunslet, Leeds; the photograph is by Barry Finch.

Martin's Cultivator Company Ltd., of Stamford, Lincolnshire, had two of their 28 h.p. tractors on view. These had four cylinders and chain-driven transmission. They weighed about two tons and one was of the crawler type. The plough was incorporated in the tractor and fitted with a power lift operated by pedal control. The price, complete with plough, was £400. The other model was a four-wheeled type and some seven cwt. heavier, the price being £450.

A rather pleasing little motor plough emanated from Mr. J. W. Maskell of Tillingham, nr. Southminster, Essex; it was later made by Messrs. Petters of Yeovil. Of great simplicity, this was the smallest type of two-furrow plough, and there were only two wheels, both of which ran on the land, the off-side wheel was the "driver" and the near-side wheel a mere balancer to keep the plough upright. The operator walked behind and held the handles as if driving horses. For its one ton six cwt., it was easily manipulated and well balanced. The price complete with plough was, however, £400, and the name of this tractor was Maskell.

The American Moline was for some time as well known as any

other motor plough in the United Kingdom. She had four cylinders, spur drive, and a belt pulley. Not unlike the Crawley, this American machine incorporated a three-furrow plough, the two driving-wheels being in front and two supporting small wheels taking the weight of the plough. There was a seat fitted on this tractor and a steering-wheel extended from the front driving-wheels. The tractor sold at £450 and there were three models present, all rated at 25 h.p., and marketed by Motrac Engineering Company, of Hazlett House, London.

The Omnitractor was a 35 h.p. twin-cylinder model from the Omnitractor Syndicate, Great St. Helens, London. Spur-wheel drive and a belt pulley made her a very strong tractor, but she weighed three and a half tons and sold at £650. Mr. Ernest Talbot supplied the picture; and tells me he took over the manufacture of this tractor in 1916.

The American Overtime had by now been established in Britain and was well advertised. There were only two cylinders, which developed 28 h.p. The transmission was by exposed teeth gearing and the tractor weighed two tons. Priced at £368, she was marketed from the Minories, London, in this country. The engine, radiator, and fuel tank of the Overtime were all independently mounted upon a main frame; these very obvious features always marked this machine and caused it to stand apart from the crowd, so to speak.

As the photograph indicates, the Pick tractor from Stamford, Lincolnshire, had a really modern appearance. There were four cylinders and the final transmission drive to the rear wheels was by roller chain and spur wheels. Weighing two tons complete with a four-furrow self-lift plough, she sold at £550. Her horsepower rating was 30.

One machine which was entered, but did not participate in the trials, was the Santler, from Malvern. Only two cylinders produced 25 h.p. Spur-geared transmission was fitted, but there was no belt pulley. It was one of the self-contained motor ploughs complete with two sets of ploughs, one set for ploughing one way, and the other for use when travelling in the opposite direction. Of the four wheels, two were in the centre of the machine and one at each extreme end supporting the ploughs.

The Saunderson Universal is described in the preceding chapter. By 1919, two cylinders were incorporated and the tractor was rated at 25 h.p. Spur gears were still fitted, but the price had risen to £510. At the trials there were three models present and the judges described this tractor as "a good general-purpose tractor—substantially built, simple in construction, and easy to handle and turn at the headlands". Accessibility was also a good

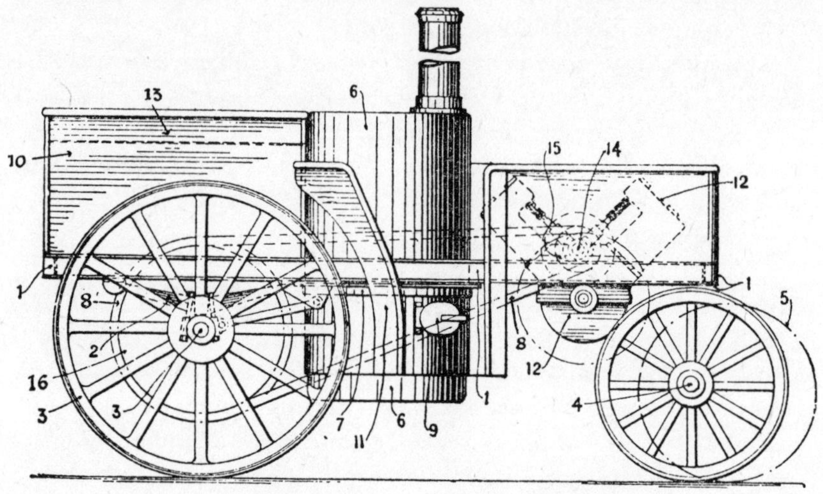

Fig. 3. Side sectional view of Summerscales' tractor

1. Girder frame
2. Spring
3 and 4. Front and rear axles
5. Single front wheel for steering
6. Super-heated vertical boiler
7. Encircling bands
8. Renold chain
9. Showing side-firing arrangement
10. Water tank
11. Flange
12. Four-cylinder 25 h.p. V-type engine
13. Coal bunkers
14. Camshaft to operate valves
15 and 16. Chain wheels

feature, and both engine and radiator, as well as the gearbox, could be dismantled without disturbing the rest of the assembly.

A firm called Summerscales Ltd., of Keighley, Yorkshire, had a 25 h.p. Summerscales four-cylinder steam tractor at work. Chain-driven transmission and a belt pulley were featured. She weighed four tons and cost £600—the only three-wheeled steam tractor on record.

The Ancone Motor Company were showing three Wallis

Junior tractors, each of 30 h.p. they were a Massey-Harris project from Racine, Wisconsin. Each had four cylinders and weighed one and a half tons. The gearing was totally enclosed down to a live axle and the price was £420. The bearings all ran in oil and this little three-wheeler was extremely popular.

W. Weeks & Son, of Maidstone, made the Simplex, a four-cylinder tractor of 22 h.p. Transmission by toothed gearing gave three forward speeds, and the tractor had a car-like appearance and weighed twenty-five cwt. A belt pulley was fitted and this little tractor was supplied with rubber tyres for road haulage purposes. With its top gear it was quite a fast machine. Priced at £400, the spare wheels fitted with rubbers cost £70 extra.

The last entry in the trials, alphabetically, referred to the Whiting-Bull tractor, a twin-cylinder American machine of 24 h.p. weighing two and a half tons, bevel-geared drive, and with a belt pulley, which cost £395. It was designed as a one-man tractor to work with a self-lift plough, and this was by no means common in 1919. Two of these tractors took part in the trials.

IV

AMERICAN INFLUENCE

As in England, the early development of the tractor in the U.S.A. followed the pattern of the steam traction engine, i.e. the stationary engine was mounted on wheels, and subsequently a drive was devised to make it self-propelled. As early as 1889 the Burger tractor appeared, built by the Charter Gas Engine Company. The chassis bore a close resemblance to a steam traction engine, into which a single-cylinder Charter engine had been mounted. The designers were obviously of the opinion that the tractor driver would feel happier sitting behind a typical steam chassis—the sort of thing he would be quite at home driving, rather than if the whole project had been radically changed. The gears were all open, of course, and mud and dirt (never absent from any form of farm work) simply played havoc with cast-iron wheels, which were not at all easy to lubricate. The gears quickly wore out, in spite of the driver's great efforts to tend the primitive greasing-cups and wick-feed oil pockets which were the only lubrication given to the main bearings. The piston was equipped with "sight-drip-oilers" and the same lubrication even dealt with the connecting-rod bearings. The driver only had hard grease and a scraper with which he tried to inject lubricant into the teeth of the pinions. Nevertheless, between 1890 and 1895 the firm built six more of these machines to the order of farmers in the north-west. Evidently these men felt there was an improvement from steam power in their particular instance.

In 1890 George Taylor, of Vancouver, patented a walking-type of motor plough, using a petrol engine. This must easily have been the first of a line of garden tractors leading up to those of the 1960s. A year later William Deering & Company made a 6 h.p. twin tractor, which in time was followed by more powerful models.

In 1892 the Case Company built their first tractor powered by a twin-cylinder opposed engine, and here again there was the typical traction-engine appearance. The greatest defect in this model was undoubtedly the ignition, which, being worked by a piston, left a lot to be desired in timing.

More attention than ever was now being given to petrol as a farm fuel; some petrol tractors were built and even rejected before reaching the market. Amazing exceptions, however, were tractors like the Otto, built in 1894 and followed by thirteen other models made and sold up to 1896. The basic design was still the steam engine, and Van Duzen in 1894 built the prototype of the Huber line of tractors, and made it so like a steam engine that he even provided a whistle. S. S. Morton of the Ohio Manufacturing Company built a friction-drive chassis with a hopper-cooled engine, and this would appear to be quite the earliest tractor with a chassis specially designed to take a petrol engine. This same company at a later date built tractors for the International Harvester Company.

The Hart-Parr tractor was interesting and had a heavy engine which was nevertheless air cooled. The makers claimed that one of these original models was working on an Iowa farm for seventeen years! By 1903 this firm was building really good tractors. Between 1902 and 1908, in fact, Massey-Harris, Lauson, and International Harvester were all in the market. The first of the last-named make built in 1905 was a small 10 h.p. friction-drive model with a reversing gear; soon there were 200 single-cylinder examples of these on the market. In 1908 the public were able to see the first trials at Winnipeg, which were repeated each year until 1912. With few exceptions, the tractors of this period were four-stroke, had automatic intake valves, hit-and-miss governors for speed control, and make-and-break ignition systems. As a rule, the electric current was from dry batteries for starting, and a low voltage magneto or generator furnished the subsequent current.

The frames of the tractors were built up on channel-irons to which the engine was bolted. Most large transmission gears were of cast iron and a wide variety of clutch types were provided.

Henry Ford's first tractor should be mentioned here. It was a 24 h.p. four-cylinder vertical engine with a copper water-jacket, vibrator coils and high-tension ignition. Rear wheels were those taken from an old self-binder! Front wheels, axle, steering-column and radiator were taken from the model K Ford Car. The drive was by spur wheel and this is noteworthy about a time when chain drive was rather commonly the accepted method. It was built in 1907.

A significant development in 1913 was the introduction of the Bull tractor from Minneapolis. This was the forerunner of the small units towards which practically all the manufacturers were now turning at this time. A delightful contemporary picture of the 1914 Steel-King shows a large gig umbrella as a permanent fixture over the driver's seat. As in Britain, the manufacture of American tractors was accelerated by the first world war, and the advent of smaller and more versatile tractors. By 1918 there was also a shortage of horses and men on American farms, and in that year alone 132,000 tractors were produced. Henry Ford had announced that his Fordsons would only be sold to State and National Governments. These went out in allotments of 7,000 and the distribution in the U.S.A. through Ford car dealers' agencies was a wonderful step forward; by August 1920 Ford claimed to have sold 100,000 tractors.

An important development of 1918 was the introduction by International Harvester of a practicable power take-off mechanism on its tractors. This permitted operating and controlling mounted and drawn equipment by the tractor's own engine through a special shaft and under the complete control of the operator. Quite soon all the leading makes were thus equipped. Although solid rubber tyres were used on industrial tractors in the early 1920s, they were unsuitable for farm tractors, because they provided little or no traction on grassy or muddy fields. It was therefore inevitable that the industry looked to the rough-treaded pneumatic tyres for the solution to their problems. In 1932 Firestone Tyre & Rubber Co. made a specially designed tyre which was adopted by Allis-Chalmers, who introduced them to their tractors and inaugurated a trend which literally put the tractor industry on air.

The year 1920 marked the culmination of a continuous upward trend from 2,000 in 1909 to 203,207 in 1920. It would be beyond the scope of this chapter to list every known make, and in other

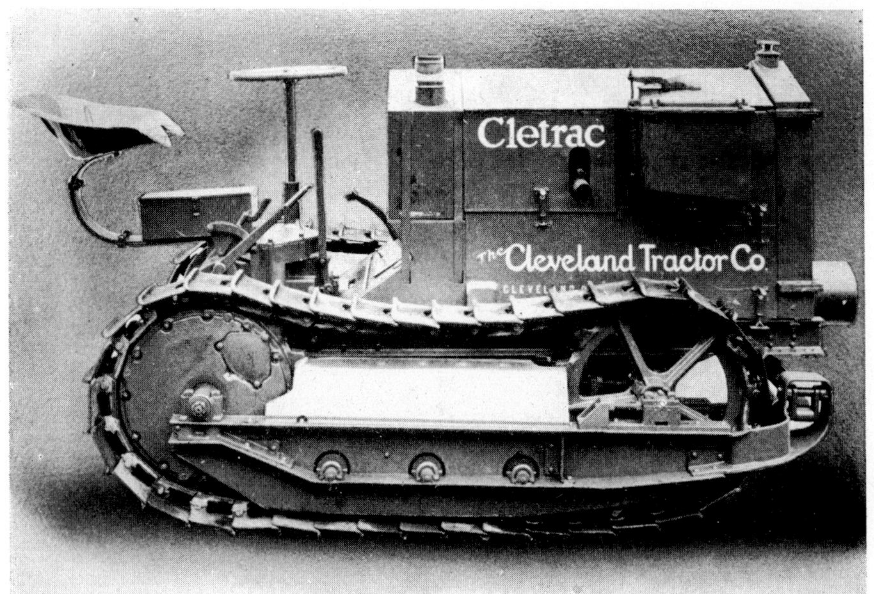

The Cleveland from Ohio.

Blackstone's little track-layer.

Alf Pepper demonstrates the Gyro-tiller.

Fowler's Gyro-tiller, another viewpoint.

1920 Crawley motor-plough.

1917 Moline motor-plough.

1915 Titan still used.

1917 Titan iron horse.

chapters I have tried to mention all those which found their way into Britain. The following table, however, shows the large number of companies manufacturing tractors in that inclusive period:

1904– 6	companies	1913– 39	companies
1905– 6	„	1914– 58	„
1906– 9	„	1915– 61	„
1907– 8	„	1916–114	„
1908– 6	„	1917–124	„
1909– 9	„	1918–142	„
1910–15	„	1919–164	„
1911–20	„	1920–166	„
1912–31	„		

By 1920 the introduction of the following list of refinements was apparent—one-piece framing, replaceable wearing parts, force-feed and pressure-gun lubrication, enclosed transmission, carburettor manifolding, air cleaners, electric lighting and starting. There was also the high-tension magneto ignition with impulse-starter, enclosed cooling systems, anti-friction bearings, alloy and heat-treated steels and the power take-off. A lightweight low-priced tractor was widely accepted. Production between 1920 and 1924 fell, largely because of the post-war depression. Keen competition brought an improved-quality tractor, but forced many companies out of business, and the same thing was happening in England. The motor plough or motor cultivator went out in favour of a general-purpose unit which with proper attachments could do all farm work. Some sixty-nine makes (some in duplicate) took part in trials at Lincoln, Nebraska, and as in Britain, the continuance of such trials speeded improved designs and eliminated many inferior makes of tractor. I append a list of those makes taking part in the 1920 Nebraska trials: Avery, Avery-Culti, Allis-Chalmers, All-Work, Aultman-Taylor, Bates-Mule, Beeman, Case, Cletrac, Coleman, Emerson-Brantingham, Flour-City, Fordson, Frick, Gray, Huber-Light, Heider, Holt, Hart-Parr, International, Indiana, Lauson, La Cross, Minneapolis, Moline, Monarch, Oil-Pull, Parrett, Port Huron, Samson, Square Turn, Titan, Twin-City, Toro, Townsend, Uncle Sam, Wallis, Waterloo Boy, and Wisconsin.

Production increased from 25,000 in 1922 to over 30,000 in 1923. Changes were largely individualistic in character and there was a move toward improved engineering design and practice. The use of anti-friction bearings made a great headway and there was greater use of ball-bearings and roller-bearings. Enclosed transmissions were designed to exclude dust.

Only a dozen makes were tested at Nebraska in 1923 and they were: Avery, Avery Track Runner, Best (two models), Bear, Case (two models), Hart-Parr, McCormick-Deering, Russell Little-Boss, Russell Big-Boss, and Wallis. A year later John-Deere, Oil-Pull, and Monarch were added to the list. The year 1925 saw some eight tractors tested, these being: Farmall, Heider, Minneapolis, Monarch, Oil-Pull M, Oil-Pull L, and Oil-Pull R, together with the Shawnee. In January 1927 some 5,000 American tractors were shipped to Russia.

Before touching on the Henry Ford story, it is worth while looking back and realizing that the very word "tractor" meant little or nothing to the general public until the 1906–7 year, when it first appeared. That year was the time in which manufacturers' advertisements used the term as a brief way of saying "petrol or gasoline traction engine". Clumsy and heavy by today's standards, these new internal-combustion engines were lighter and more nimble than their steam forerunners. In 1908 a single-cylinder machine was completed, weighing 10,000 lb. Today a tractor developing more than 50 h.p. will only weigh 7,000 lb. The early track-layers laid their own rails, so to speak, in the form of shoes made of wooden blocks. These were, of course, the forebears of present-day crawlers. The International Company built their first tractor in 1906, and two years later, competing in a demonstration at Winnipeg, took second place to a Kinnard-Haines model. At that time, Canada was the best customer and took two-thirds of the U.S.A.-built machines. The Rotary Hoe was produced commercially in 1912, and it was soon found that these worked best at a high speed behind tractors rather than horses. Garden tractors came into the picture in 1916 and certainly lightened the burden of smallholders. The tricycle-type tractor combined draw-bar usefulness with the ability to do row-crop work, a very great feature in today's farming.

By 1928 Massey-Harris of Racine had taken over the J. I. Case works in the same town, thus giving themselves a famous line of farm implements, including the Wallis tractor. Case had bought the Emerson-Brantingham Corporation, which had been making farm equipment in unbroken sequence since 1852. By 1928 the Caterpillar Company was producing a 25 h.p. model and many other track-layers had been greatly improved. The following table gives the number of manufacturers from 1921 to 1930; the decline in their number which I have already mentioned is apparent:

1921–186	makers	1926–69	makers
1922–116	,,	1927–61	,,
1923– 93	,,	1928–51	,,
1924– 64	,,	1929–47	,,
1925– 58	,,	1930–38	,,

At the Nebraska trials of 1930 only ten tractors appeared. They were: Cletrac, Eagle, Fordson (two models), Massey-Harris, Monarch, Oliver Hart-Parr (two models), Oliver Row-Crop, and Twin City. By this time Massey-Harris were making a four-wheel-drive tractor.

Ford's biggest rivals would not grudge my devoting the conclusion of this chapter to the Fordson story. Henry Ford was the son of a farmer, and even before he produced a car, he had planned a farm tractor; mention has already been made of his first effort of 1907. He gained his mechanical knowledge by repairing traction engines as used for steam ploughing. He realized, however, that the future lay in a smaller machine to carry its own implements. His maxim was that the power unit must be integrated with the tractor.

On March 8th, 1911, the English-made Ford came into being when the company opened up in Shaftesbury Avenue, London, with a capital of £1,000. The first managing director was Mr. Percival L. Perry (later to become Lord Perry), who with a dozen employees set about assembling imported parts of the famous model T. From such a modest start has grown the largest single vehicle-producing firm in Britain.

The declaration of war in 1914 thrust Perry and not Ford into the limelight, for the simple reason that the first Henry Ford was a great pacifist. Perry offered the entire productive capacity of the works to the British Government. Cars, ambulances, trucks, and munitions were produced. In 1916 Perry was advising the Food Production Department and in 1917 became Director of Agricultural Machinery, an appointment which led to the development of the Fordson tractor and the fuller mechanization of British farming. The German submarine blockade had by 1917 reduced food supplies to about two weeks only. The Prime Minister (David Lloyd George) called in Perry and gave him a free hand to speed up home production. Horses had been commandeered by the Army and many young farmers were called up. Mechanization was, of course, the only answer, and 5,000 tractors were shipped from the U.S.A. I give a picture of the first model and it is amazing to notice how very little the general design has had to be changed since. A further 1,000 were soon sent, and by five months the 6,000 had arrived. The model was well in advance of many experimental tractors. They were mounted on steel wheels, had a worm and wheel final drive, and a water air cleaner. Petrol or paraffin could be used and the engine developed 23 h.p. on petrol and about 20 h.p. on paraffin.

Perry persuaded Ford to build a tractor factory in Cork and Henry quickly agreed, for Ireland was, in fact, the home of his ancestors, and he felt he could aid the war effort without fully surrendering his pacifist beliefs. The Fordson had played a really wonderful part in U.S.A. tractor development from 1918 to 1928. Starting with a production of over 34,000 in 1918, it is amazing to realize that this figure represents over a quarter of the 132,700 tractors produced by all the other 142 American companies. For the next two years the Fordson enjoyed a share of the increasing business with 100 per cent increase over its 1918 figures. When in 1921 the post-war depression was being felt, the low-priced Fordson represented about 50 per cent of the total tractor population. In 1923 and 1925 over 100,000 Fordsons were produced annually; this was 75 per cent of tractors produced by all other companies. Production of Fordsons in America ceased in 1928 until 1939, when the Ford-Ferguson was introduced. When the plant was moved to Cork opportunity was taken to redesign the tractor, and the bore was increased from 4 to $4\frac{1}{8}$ in.; a high-tension ignition system replaced the low-tension flywheel magneto with its somewhat troublesome vibrating coils. The thermo-siphon cooling system was replaced by pump circulation. The transmission was more heavily built, and taper roller-bearings were now used. The front end was spring mounted and fenders became standard equipment.

I have often discussed with farmers and their men the phenomenal success of the Fordson. It is worth noticing that the Fordson's weight was only one cwt. per horsepower compared with three or four cwt. per horsepower of many other makes on the early market. Weight, after all, is only needed to aid adhesion and to give strength. By combining good materials and workmanship, Ford succeeded where others failed. Farmers in Suffolk and Essex when I was a boy used to say that no tractor should exceed thirty cwt. to be a success on our clay soils, and even then, with the early wheels, there were times when the tractor was best in the cart shed, especially during wet weather.

One look at the early Fordson reveals how easily heavy and costly impedimenta could be dispensed with without impairing efficiency, the most noteworthy example being the entire absence of the usual frame with its cross-members and brackets. The machine in reality was a power unit and transmission gear *en bloc* mounted on two pairs of wheels, together with the simplest control and steering arrangements. The whole of the transmission was enclosed to run in oil, and yet there were sixteen places for the operator to oil by hand; all the other parts were oiled automatically.

In a very fine little book on the life of Christ, *The Man Nobody Knows*, the author, Bruce Barton, an American journalist, has a vivid memory of an interview with Henry Ford. I will quote it, beginning with Ford's opening remark:

"Have you ever noticed that a man who starts out in life with a determination to make money never makes very much?" asked Ford. Without waiting for comment he went on to answer his own question. "He may gather together a competence, of course, but he'll never amass a huge fortune. Let a man start out in life to build something better and sell it cheaper than it has ever been built or sold before—let him have *that* determination and the money will roll in.

"When we were building our original model, do you suppose that it was money we were thinking about? We wanted to make a car so cheap that every family in the U.S.A. could afford to have one. So we worked morning, noon, and night, until our muscles ached and our nerves were so ragged that it seemed as though we just couldn't stand it. . . . One night we were almost at breaking-point and I said to the boys, 'Well, there's one consolation, nobody can take this business away from us unless he's willing to work harder than we have worked'; and so far," he concluded with a whimsical smile, "nobody has been willing to do that."

Herein, surely, lies the success story of the Fordson tractor.

V

THE TRACTOR MECHANISM

THE trials which were held substantially helped the farmers to choose their type and make of tractor. The choice, however, must have been bewildering indeed, with so many makes on the market. On the other hand, many of the smaller differences in design did not greatly influence the work these early models performed. The differences which really mattered were more of weight, horse-power, width, and the general disposition of transmission and wheels.

In the latter case, for almost any other job than ploughing, it was usually considered that the two-track four-wheel machines were best. Gradually these types superseded the other ones. Wheel grip was of very great importance and here on heavy land twelve-to-fourteen-inch-wide wheels were best. Often such a tractor ran with both driving-wheels on unploughed land. In some cases the front wheels were set closer together than the driving-wheels, making it impossible to run in any width of furrow. On East Anglian clay land, I know these tractors often came adrift altogether. There certainly was occasion then for the sceptical farmer's charge of "packing" the land. Each subsequent operation in preparing the seed-bed was rendered more difficult.

It will thus be seen that the advantages of wide wheels could be sometimes outweighed by the disadvantage of the extra weight all being on unploughed land. Thus it was that the two-track four-wheeled tractor with two narrower driving-wheels (ten inches wide) became popular. With this type, using a ten-inch plough, it was possible to run the two off-side wheels in the furrow so that soil-compression was halved. Because of the narrowness of the wheels, a certain amount of adhesion was sacrificed, but with less compressed soil the same power was not required to turn it up. There was one other advantage in that with only one wheel running on unploughed land the surface was not chopped up to the same extent by spuds or strakes as was the case with the two wider wheels. The work could be "laid" pretty evenly without using ploughs with too long breasts.

Many farmers whose land was unduly heavy were attracted by chain-track and two-track three-wheeled models. In the case of the latter, a three-wheeled machine, having one large driving-wheel, had an idler (or small driven wheel) running on the unploughed land merely to preserve the balance and carrying very little weight. With such a tractor, practically all the weight was at the bottom of the furrow, as with a team of horses. Such a tractor did not, however, possess the same adhesive qualities as the one with two large driving-wheels. Often it could work on wet land when no other tractor could run at all; this was because with a dry bottom to the furrow a wet surface was less important, as all the driving was done by the big wheel. Tractors which ran with all their wheels out of the furrow were very difficult to steer, and, to aid steering, some makers supplied an automatic steering-device. This consisted of a small furrow wheel attached as an outrider to the off-side front wheel. The machines which ran simply with their off-side wheels in the furrow were almost self-steering, and as a schoolboy I have watched their drivers walk by the side of the plough in cold weather. Such tractors were more adaptable to the use of the self-lift plough when this was first invented. A tractor with only one driving-wheel required no differential gear and less driving-gear—usually it cost less for this reason.

Engines likewise varied a great deal. There were single-cylinder, double-, and four-cylinder vertical engines, and tractors with single- and double-cylinder horizontal engines. These I have tried to illustrate in diagrams. By about 1920 the commonest types were either the twin-cylinder horizontal engine, which developed its maximum power at about 500 revolutions per minute, or the four-cylinder vertical engine developing its full power at about 1,000 r.p.m. Each was at that time about equally popular.

The makers of four-cylinder vertical engines claimed that these were lighter, freer from vibration and more efficient than twin-cylinder horizontal engines of equal power. The makers of the twin-cylinder horizontals, on the other hand, claimed that less power was absorbed by the reducing gear on account of the lower

The 1921 Mann steam-tractor at Woodford Bridge Rally 1959.

Fowler's two-wheeled motor-plough.

The Omnitractor of 1916.

Samson's three-wheeled "Sieve-Grip".

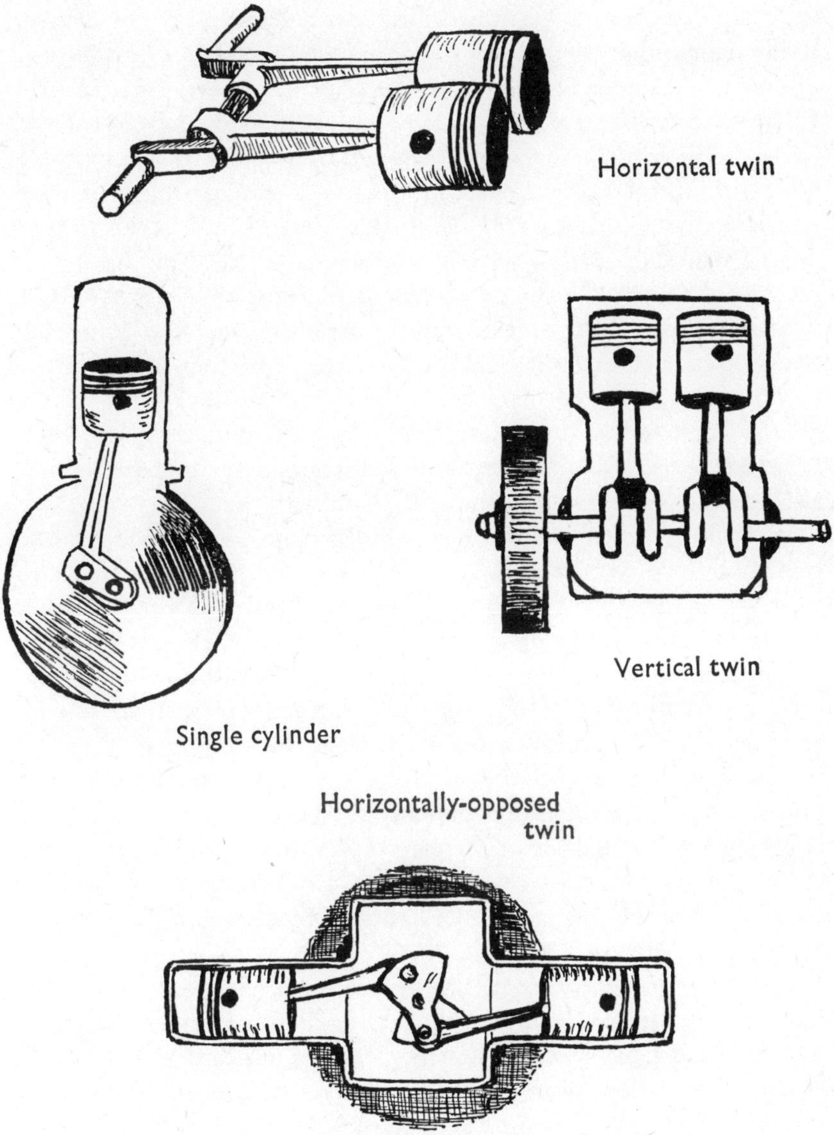

Horizontal twin

Vertical twin

Single cylinder

Horizontally-opposed twin

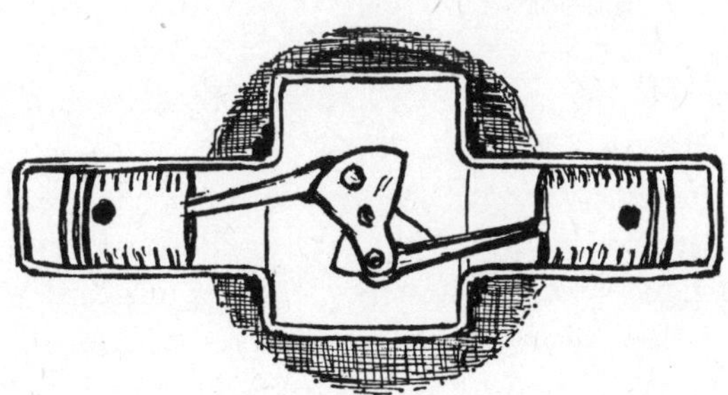

Fig. 4. Types of engine

speed, and because the engine lay across the frame with the crankshaft at right-angles to the direction of travel, thus dispensing with the need for bevel and worm gearing. Here, too, were simplicity and fewer working parts. Gradually, of course, the four-cylinder engine was to win its way to greater popularity.

During 1922–24 I was a student at Chadacre Agricultural Institute and was fascinated by the lectures then given by Mr. Peter Syme, B.Sc., on the working of the internal-combustion engine. I have from time to time passed on this simple rudimentary lesson to young farmers and others. It is amazing even in the 1960s to find people who habitually use these engines, in either tractor, car, or motor-cycle, and yet do not really know how they work. Those who do understand will obviously wish to skip the following paragraphs.

The common type of tractor engine, as we have seen, was a four-stoke, whether using one, two, or four cylinders; the Ivel-Hart, however, was a two-stroke. I will try to describe both actions as I learned them in 1924, and will begin with the four-stroke cycle of operations. In this class the engine develops one working stroke in every four. With the suction stroke, the movement of the cylinder coincides with the opening of the inlet valve, and a mixture of air and gas is drawn into the cylinder. Then follows the compression stroke, the inlet valve closes and all the gas is compressed into the combustion chamber. The third is the expansion or working stroke in which the compressed gas is ignited by a spark and the resulting explosion drives the piston down. The momentum thus generated travels to the flywheel and is given out in the form of power. The completing stroke is the scavenging or exhaust stroke, when the piston returns, the exhaust valve opens and the spent gases caused by the explosion escape through the open outlet.

The two-stroke cycle was less developed for tractors and today is far more common in small motor-cycle and lawn-mower engines. Yet this principle was the first to be invented and a number of tractors were thus fitted. The two-stroke engine, of course, has greater simplicity. Economy is effected by the fact that there is no exhaust stroke, the foul gas being allowed to escape at the end of the working stroke through a port uncovered by the piston. The compression stroke is also used to draw a fresh charge into the crankcase. Thus it comes about that with this type of engine there is one explosion per revolution, i.e. one explosion for two strokes of the piston. Since no valves are required to operate such an engine, there are fewer working parts, price could be lower, and in those early days the engine was extremely reliable. It could be argued also that since it gives one explosion to one revolution,

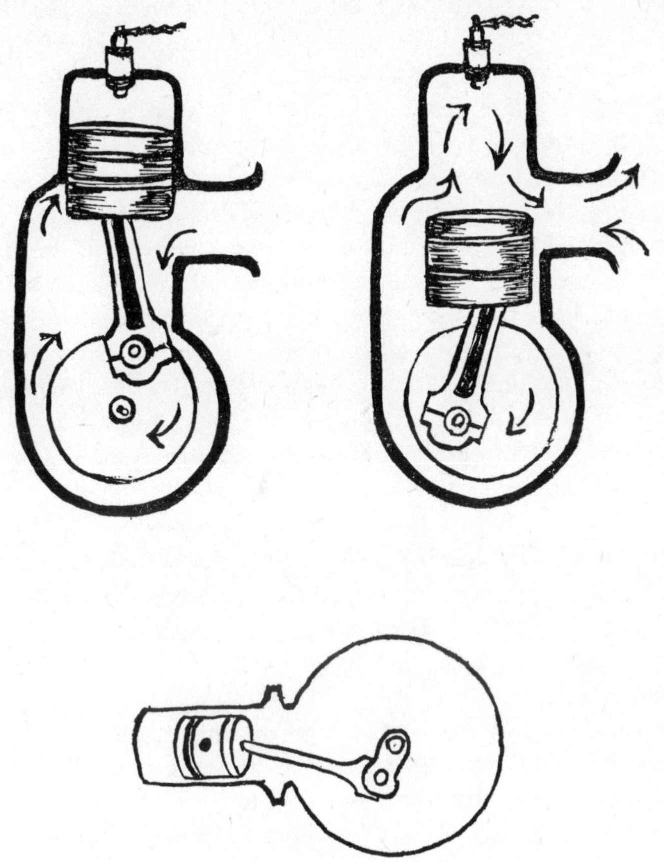

Fig. 5. Two stroke principles

as against one to two in a four-stroke, a two-stroke ought to develop twice the power. This did not necessarily follow in every case. Owing to the greater number of working strokes, a smoother-running engine resulted. A disadvantage was that owing to the greater number of explosions these early engines were only efficient at low horsepower ratings. Complete scavenging and perfect lubrication were also difficult. With more people using tractors for belt work (stationary), it was soon realized that farm tractors needed some sort of governing device to keep the engine running at a constant speed independent of variation in the load from time to time. Many systems were tried out, but they can be divided into two main classes, namely, the "hit or miss" and the "quality of mixture". Again we see the steam traction engine influence, for the actuating mechanism consisted of a weight working on a central shaft driven by gears from the timing-shaft.

In the "hit or miss" type the cylinder receives the full charge of air and gas at each suction stroke, or it receives no charge at all. In the event of the speed tending to become too high, the exhaust valve is prevented from closing and the following suction stroke simply draws air back through the exhaust; this meant a working stroke was being omitted. Thus there would be eight idle strokes between explosions. This entailed a fairly wide variation in speed.

In the "quality of mixture" instance the speed is varied by altering the amount of fuel admitted to the cylinder during the suction stroke, but the amount of air is unaltered. The objection to such a method is that the proportion of air to gas cannot always be, say, ten to one, which would be economical. On the other hand, since the engine gives one power impulse for every four strokes, it means slight variation in speed, and an even "take-up" of running.

Until the wider adoption of diesel oil for farm tractors in these modern days paraffin was almost exclusively used. But one disadvantage was inability to start a cold engine without a little petrol, and a two-way tap was provided on most models. Carburettors seem to have changed very little in the course of the years. Paraffin does not evaporate at ordinary atmospheric temperatures, so before being converted into explosive gas it needed to be heated. Quite soon, therefore, the early manufacturers utilized the hot waste gases passing through the exhaust pipe, which was done with the aid of a vaporizer. This consisted of a chamber (through which hot exhaust gas is passing) surrounding the induction pipe. Sometimes the air was heated before reaching the carburettor by drawing it through a chamber surrounding the exhaust pipe, and one or two makers coiled the feed pipe around the exhaust pipe. Some tractors ran much better on paraffin than others, and one heard of cases where farmers had to substitute the more expensive petrol to get any sort of result at all.

As compression causes heat, so expansion reduces temperature; therefore as water becomes vaporized it has the effect of drawing off some of the explosive heat of the charge before it is compressed. A water-drip system was therefore introduced (see sketch). It was really a small pipe fitted with a tap leading from the cooling system to the induction pipe. The tap had to be carefully regulated, and when the tractor was idling it could indeed be turned

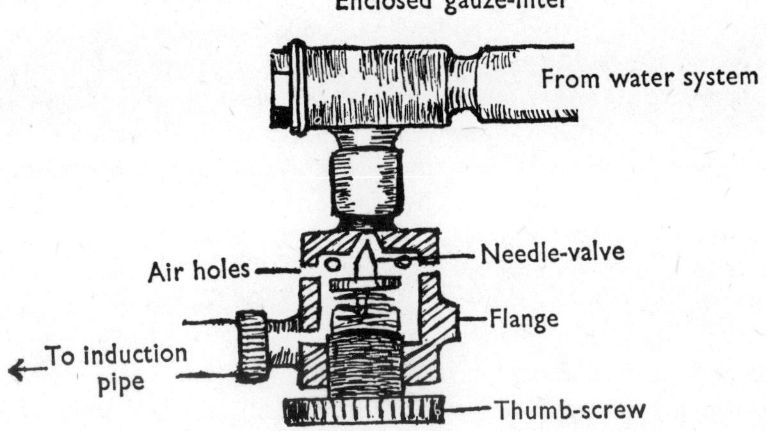

Fig. 6. The water-drip cooling system

off completely. The one sketched shows a spring-controlled valve to regulate the supply of water.

Field work is always dirty, and in dry weather dust and grit are flying about. To guard against such solids being drawn into the air intake for the engine some sort of filter had to be provided. Sometimes it was a tin box stuffed with wool from a sheep's back. Sometimes air was drawn in from a long chimney which stood well above the tractor body where the air was clean. A really excellent method adopted by some makers and mentioned in another chapter was to draw the air through a tank of water. Such a plan moistened the air so that the water-drip was rendered unnecessary.

Farmers have always been accused of being sparing with the oil-can, and of leaving valuable implements outside in all weathers. Many discovered to their cost that this neglect just would not do in the case of tractors. There were several methods of engine lubrication, and these were automatic. The driver, of course, had to make sure the reservoir was charged with oil, and sometimes a sight-feed was fitted; others had a safety level. My first introduction to this tap was on the early Fordson when I put in some months' training with Mann Egerton of Bury St. Edmunds.

In most systems the oil held in the reservoir was pumped mechanically to the various bearings. The oil in some cases was forced up the oil pipes and along the drilled apertures to the main and big-end bearings. From here the oil found its way back through the hollow crankshaft to the connecting-rod bearings. In another case, the oil was pumped into narrow troughs, from which a scooped dipper on the end of the connecting rod splashed sufficient oil to lubricate the bearings, as well as the walls of the cylinder. There were many spots on these old models which needed frequent oil and greasing.

Tractor engines had, of course, to be cooled, and although some very early models were designed for air cooling, it need hardly be said that such a system—well enough for a fast-running motor cycle or aeroplane—was not the sort of thing for a slow-moving tractor. The cylinder was therefore surrounded by a water-jacket, which communicated with the radiator by top and bottom piping; a fan also helped, as in car engines. Sometimes a large tank took the place of a radiator, this being placed higher than the cylinders. As the water got hot in the jacket it rose through the top pipe into the tank, whilst the cooler water in the bottom of the tank flowed in to take its place. Such tanks may be seen in many of the illustrations which form the plates of this book. Such a system needed far more water than did a radiator and fan. A tractor of about 20 h.p. needed between twenty or thirty gallons per day when engaged in heavy ploughing. Two or three gallons would keep a radiator going all day. Thus it was common in these early days for horse transport to cart water to a tractor working away from a water supply, just as was done for steam engines! With the tank system an engine ran a lot hotter and the water-drip device described earlier in this chapter was more than ever necessary.

The ignition on early tractors was by magneto and these early ones on the whole were very reliable. The usual type for a four-cylinder engine had a reel or armature wound with two coils of insulated copper wire which revolved between the poles of a set of inverted U-shaped magnets. The contact breaker made the necessary interruptions and the usual distributor directed the current to each sparking plug in turn at the correct time. There was a hand lever for advancing or retarding the spark according to the load and when cranking-up to prevent a backfire—which could mean an injured wrist. Larger tractors were soon given an "impulse"-starter which caused a specially strong spark to be given when the engine was being turned by hand. Self-starters and lighting were first fitted to tractors as long ago as 1920.

The Fordson was equipped with a rather different system of ignition. Briefly, a combination of a set of magnets (sixteen in all) was attached to the flywheel. These revolved close to the same number of bobbins mounted on a plate attached to the cylinder casting. The bobbins were wound with a continuous strip of copper wire, and the revolving process generated a low-tension current. This was still not powerful enough to fire the plugs and so it was distributed to each of the four induction coils in turn (one for each cylinder). Distribution was effected by the commutator placed on the front of the engine and worked by the end of the valve camshaft. This revolving roller contacted four segments in turn inside a round case and the segments were connected to the induction coils by wires. Each coil consisted of a soft iron pillar around which were a primary and a secondary winding, all of which was in a sort of box. The low-tension current flowed via the primary into the secondary windings by the quick interruptions of the low-tension circuit, by means of a trembler. There was a trembler on each induction coil and their contact points, I remember, often gave trouble unless kept spotlessly clean and adjusted. It was possible to adjust by the adjustable nuts, and smoothing was sometimes necessary with a file.

Even very early tractors had some sort of clutch for gear-changing purposes. The simplest form was the friction clutch, known as the "cone clutch", too well known to need description. Some types were liable to slip badly and to be difficult to withdraw at times. Plate clutches became popular, and other forms were either the expanding clutch or the contracting clutch. In the former, metallic shoes expanded inside the hollowed flywheel, gripping the internal circumference. In the latter, metal strapping

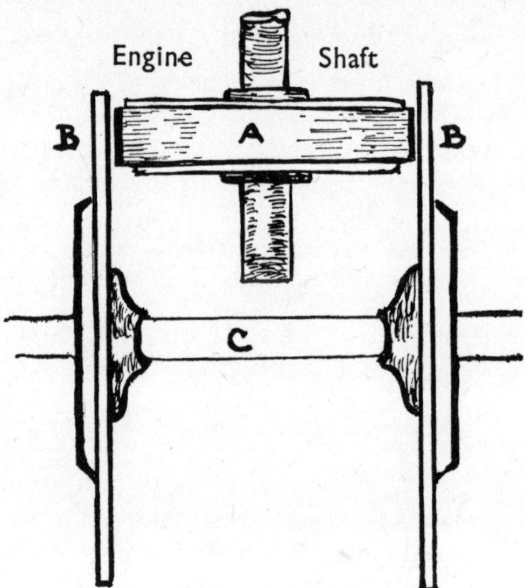

Fig. 7. Friction change-speed

contracted to grip the outside of the drum; the metal, of course, had a frictional fabric.

Rapidly coming into favour was the worm-driven gearing which I illustrate, and which gives in a small space a form of transmission running in an oil-tight casing. Chain drive was on its way out; too much exposure and interference by weather, dust, and stones frequently made it a troublesome drive. From America came a change-speed gear of frictional design. By sliding the discs along the countershaft, a middle position of neutral could be given. Reverse gearing was soon a normal fitting embodied into tractors, and the differential gear had also come to stay.

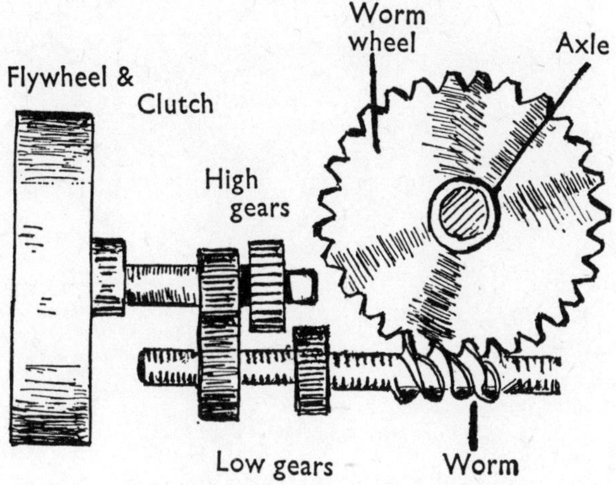

Fig. 8. Worm-wheel transmission

Steering was already divided into two varieties—hand and mechanical. The former employed a steering-column, worm and pinion with suitable rods or chains. The motor-car type of Ackermann steering was better on the lighter tractors, whilst the chain-barrel type was merely on the basis of the traction-engine arrangement of wheels mounted on the main axle, and was better for heavier machines. The axle was made to pivot in the centre, whilst the chains wound and unwound on the barrel as seen in the sketch. It is an honest criticism if we say that among the weakest points in many American tractors was the steering. Designed for big flat

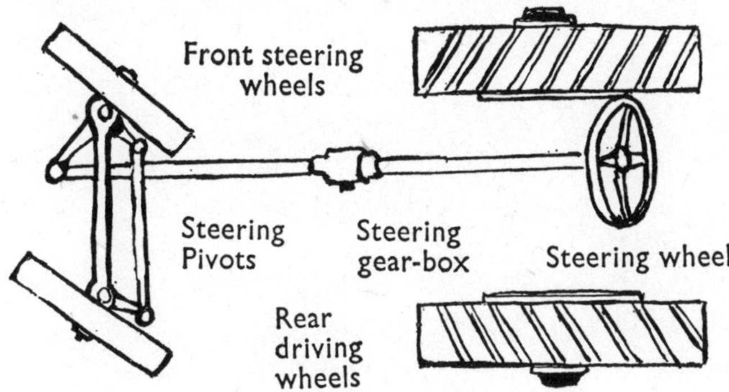

Fig. 9. Ackermann steering

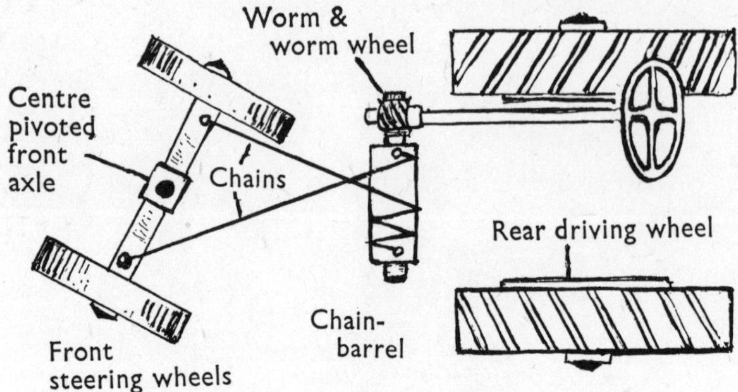

Fig. 10. Mechanical steering

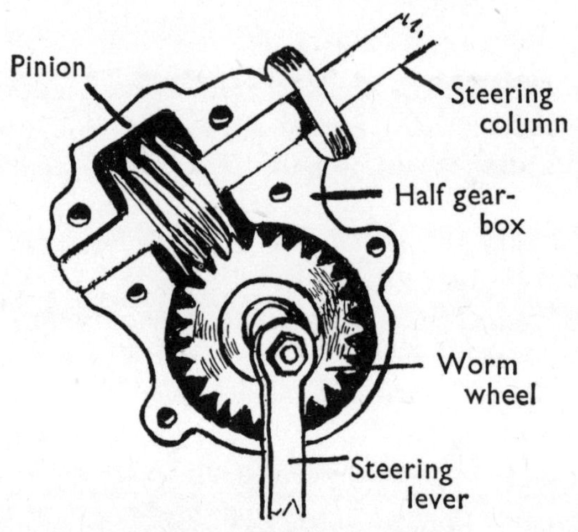

Fig. 11. Worm drive steering

open fields and ranch farming in both the U.S.A. and Canada, it came adrift in the small heavy-land fields of the average British arable farm.

In the next chapter I intend to mention some of the varied tasks which the early tractors accomplished. There will necessarily be an emphasis upon ploughs and ploughing.

VI

THE WORK THEY DID

In the last chapter we saw that the tractor itself by disposition of weight, cleating of wheels, width of wheels and horsepower, could greatly influence the quality of the work accomplished. Plough design also varied enormously, but could be divided into three distinct types:

1. The independent plough, upon which a second man rode in order to manipulate levers for setting in, pulling out, and regulating the furrow depth. A glance at some of the plates in this book reveals the large number of such ploughs in use.

2. The power-lift plough, which could be easily controlled from the tractor-driver's seat. The power which lifted the plough in and out of work acted through a clutch on one of the plough wheels attached to a cranked axle. By a crane attachment another similar lift could lift the plough.

3. The self-contained tractor and plough, in which both tractor and plough were mounted on to the same frame and balanced by the driving-wheels. They were steered from behind and the driver was usually provided with a seat. The Crawley Agrimotor was a good example. Some were walking models, where the tractor was simply a mechanical horse.

In 1920 the Royal Show was held at Darlington and a somewhat revolutionary tractor was shown by the Timesaver Company, of Colman Street, London. Called the Timesaver, this interesting machine was a tractor combined with a "one-way plough". In this a plough was mounted fore and aft of the tractor. It worked by starting at one side of a field and was able to work to and fro without turning at headlands. All furrows were thus turned in one direction and the machine steered itself. This made a reduction in the width of the headlands, and these did not need to be continuously run over. It was also claimed to be specially useful for working across a steep gradient. It is remarkable how soon machines of such promise were scrapped.

Good work not only depended upon the tractor, but also upon the driver and his ability to adjust the plough. Deep ploughing was fashionable in the 1920s. First-class adjustment gave furrow slices which were set up at a sharp angle, their width and depth being similar. Many imported ploughs took far too wide a furrow, resulting in poor ploughing. The drawbar pull necessary to haul a plough weighing seven or eight cwt. was ninety lb.

A plough weighing ten cwt. would thus require a drawbar pull of one hundred and twenty lb., which was about thirty lb. above the pulling power which was required by a lighter plough. It was

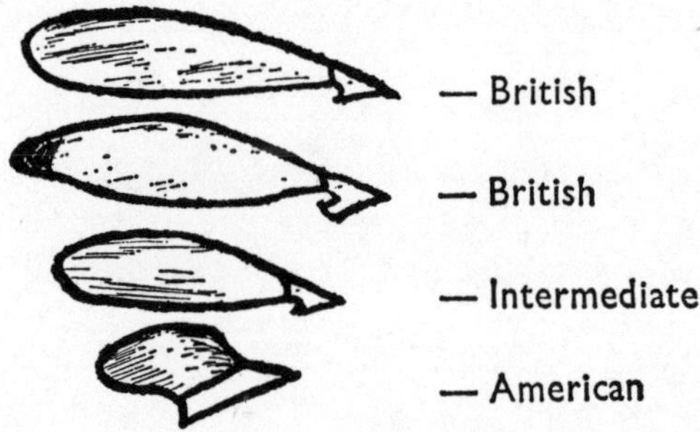

Fig. 12. Early plough types

soon learned that a heavy plough did not necessarily increase the draught. A heavy plough could, if properly constructed, be light in draught. On the other hand, a light plough could pull very heavily. It often depended upon the type of breast or mould-board, and the setting of the plough, and the way in which it was coupled to the tractor. The rough sketch depicts the four kinds of breast in general use; the two British types are similar, but one has a good deal of its underneath edge cut away. Many of the tractors hired out by the Government during the first world war used the "intermediate" size, and ploughs thus equipped pulled very lightly. Neither those, nor the American type, were ideal, however, for very heavy land, which needs a long sweeping breast, especially if grass has to be ploughed in and buried. Most breasts were held by stays or bars which could be adjusted; a tendency to set them out too far would sometimes make heavy going of the job.

Disc coulters were no new thing, but war-time tractor ploughing made them popular. They were ahead of the ordinary knife coulter, which was popular, of course, for horse-drawn ploughing. The discs cut through the grass and spear-grass which often choked an ordinary knife coulter. Good disc coulters lasted longer, too, and I suppose, by reducing draught, they could also be termed fuel-savers. In breaking up grassland—a common feature of

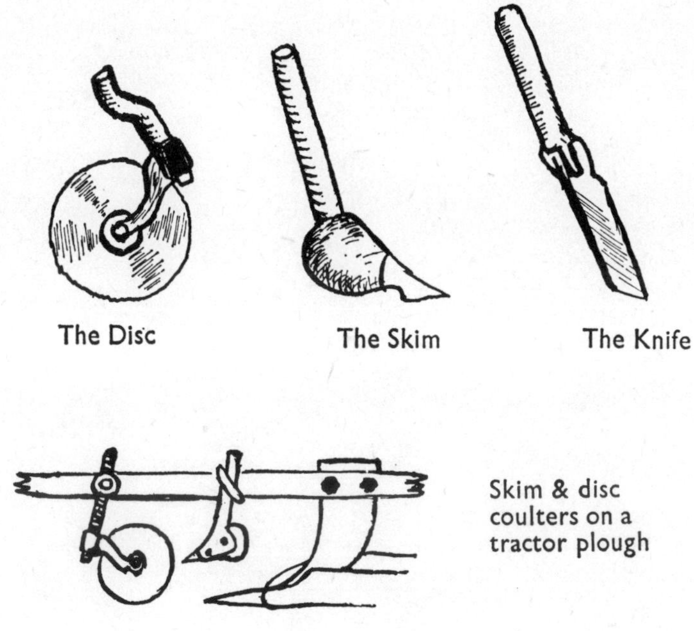

Fig. 13. Coulters

every war-time period—the addition of a skim coulter to shave off the turf was a good feature. They worked well if made to run behind the disc, as long as the latter was set well forward as in the sketch. There was a great need for better slip-couplings in those early days. Plough after plough was badly broken up by striking roots or buried rocks. There was also a need for a coupling which would give way under abnormal strain and prevent a broken plough. The simpler type I have sketched was the sort of thing a blacksmith might make. The other one incorporates a spring, and

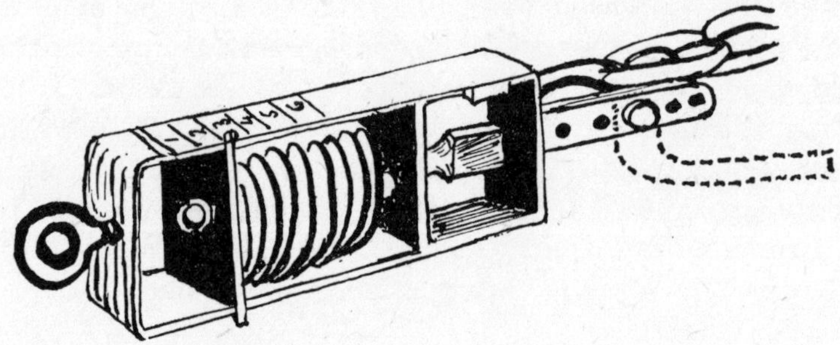

An adjustable coupling which released the draught-rope at a given pressure. Under excessive strain the hook is tipped into position shewn by dotted lines. The figures denote pre-determined drawbar-pull in tons.

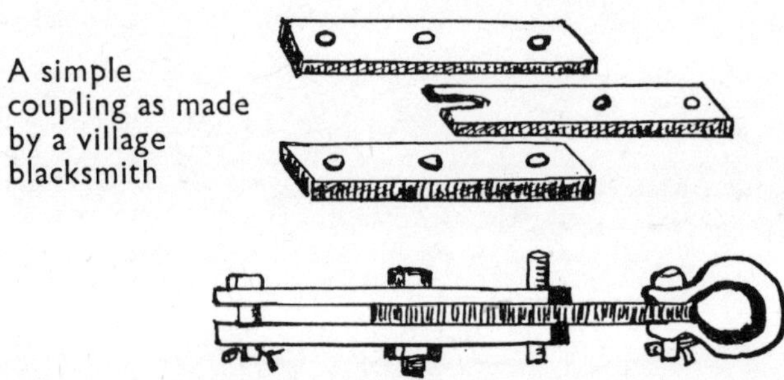

Fig. 14. Types of slip-coupling

there was a scale plate, showing in tons the sort of drawbar strain which could be predetermined. The fitting of these can easily be seen from the sketches.

In the U.S.A. and Canada the disc harrow had been in use for a great many years. This made it possible to tackle rough land where a quick seed-bed was imperative. With a double-gang harrow, one row of blades would be spade-shaped cutting types and the other row would be plain circular discs, which, being shaped like a saucer, gave the soil a side-to-side displacement. For cutting up a ploughed meadow these harrows were an ideal implement; toothed harrows would bring too much of the buried turf back again to the surface.

It is strange to recall that, certainly in Suffolk when I was a boy, it was a common thing to open out the ploughing on any field by horse plough. The tractor then came along and ploughed the field, but again the shutting-up of the furrows was left for the horsemen. How some of these ploughmen hated waiting on the tractor-drivers in this respect! Gradually the tractor plough became accustomed to doing the whole job, but through trying to leave a narrow headland many a driver has ditched his tackle. The headlands often became very compressed before they could be finally ploughed—the large tractors especially, required a wide headland upon which to turn.

For a very long time ploughing and harrowing were the only work designated for the tractor. Even when experiments were made it was usually the old horse-drawn implements which were brought in and hitched to the tractor with great, or sometimes very little, success. Cultivators, rollers, and harrows were next tried, but it was some time before the small arable farmer attempted to drill corn by tractor power. Grass-mowers and binders were tractor drawn long before the first world war. If it was only intended to draw a single grass-mower or binder, a short hitch pole was all that was necessary, and many farmers made their own hitch pole, in consultation with their local blacksmith. The firms of Whiting-Bull, McCormick, and International Harvester Company, all marketed special devices for this purpose, especially for when it was desired to tow more than one binder at a time. Overtime tractors coupled the binders step-wise, a quadrant and pinion on the hitch pole creating the necessary angle. I photographed our first Fordson at Bryers Farm towing a thistle slicer, and behind it a set of heavy duck's-foot harrows with a home-made wheeled whippletree to give the necessary lift by running between. These two jobs would normally each have had three horses and a man, on our stiff clay. Thus the tractor was doing the work of six horses and two men, even before the days of special implements. Plate 23 also shows a real mixture of ancient and modern, for the wagon was almost a hundred years old, upon which is shown a load of wheat which I loaded, and then climbed down to take the photograph.

It will have been noticed that the majority of these early tractors were fitted with a pulley for stationary belt work; threshing, wood-sawing, and baling were the obvious choices. The fixed stationary oil engine in the barn began to go out at this time, in favour of a tractor which drove shafting, which in turn drove chaff cutter, beet pulper, and corn mill. The appeal lay in the fact that the tractor could be dual purpose, i.e. a single outlay of capital covered the farmer's power plant both in the field and in the barn. Furthermore, the tractor (unless very stubborn) could be started more quickly than a stationary engine which required a blowlamp.

Threshing was the heaviest job, and it must be admitted that some of the earliest tractors did not drive the threshing drum as well as the traction engine did. An efficient governor was more than ever necessary for stiff belt work. Many of the pulleys were too small, and there was invariably a good deal of belt slipping. This was particularly true of tractors employing vertical high-speed engines where the pulley drive was taken direct from the engine shaft. A large pulley revolving at 700 to 1,000 revolutions per minute was too quick a drive; the solution was to employ a large pulley driven through a reducing gear. This reminds me of an ingenious scheme which George Oastler employed at Fyletts, Hawstead, in my schooldays. An old portable steam engine for many years had driven the millstones at this farm. When the tractor which replaced her was found too fast the old engine was brought back as a reducer. The tractor drove the steam-engine crank, and this in turn rotated the mill shafting.

A belt pulley on the early models was found in no less than three different positions; on the side of the machine, parallel to the frame; on the end of the crankshaft on the front of the tractor; or on the tail shaft at the rear. In the latter two positions it ran at right-angles to the frame. The first position was certainly the best and I have sketched the International fitting. When working this way the tractor could be driven up to its work and moved backwards or forwards until the correct tension was given to the belt. If the pulley was in position two, or position three (fore or aft),

A 1914 Mogul in use.

1916 Mogul still used.

A Moseley with special rear wheels.

Parrett tractor of 1917.

1920 British Wallis.

A Wallis Junior three-wheel tractor.

1919 Fiat still used.

1929 Rushton built at Walthamstow.

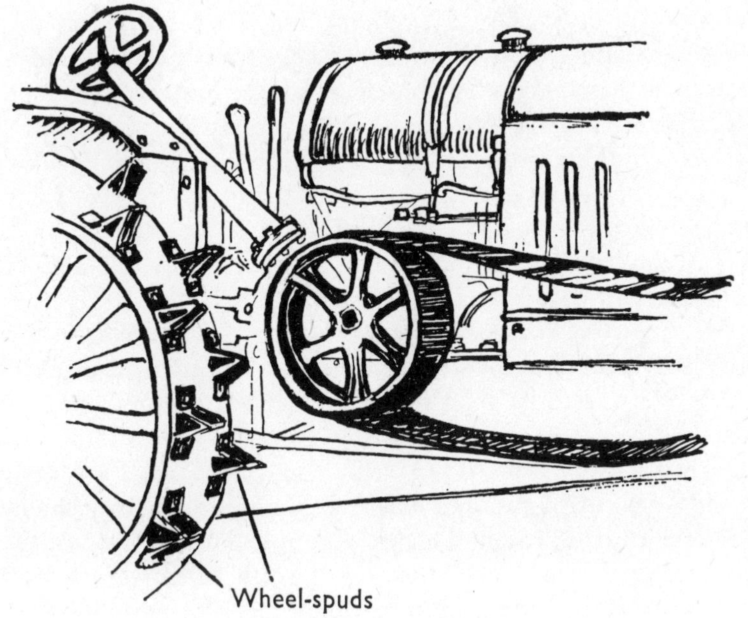

Fig. 15. The belt pulley of an early International

the tractor had to stand at right-angles to its work. This made it difficult to "set" the tractor so as to get correct belt tension, and this was not easy in barn or stackyard.

As a tractor engine could not so readily be stopped and started as a "steamer", provision had to be made for quickly disconnecting the power from whatever was being driven without actually stopping the engine. With a clutch it was simple; otherwise a pulley was used which embodied its own separate clutch.

Firms like Saundersons made their own tractor ploughs. In their catalogues they stressed the importance of the wide variety in soils which farmers should have known well enough. Great weight on the land was, of course, thoroughly objectionable, yet there were many ploughs up to 1930 which could not be pulled by anything but a heavy machine. Other famous firms to make tractor ploughs, especially in this period, were Cockshutts, Olivers, Martins, Howards of Bedford, and Ransomes of Ipswich.

Whilst preparing this work I have had a most interesting letter from Mr. A. Carleton Whitlock, of the well-known Essex firm of

Whitlock Brothers, Great Yeldham. He tells me that his grandfather, the late Walter Whitlock, of Pool House, Great Yeldham, used to farm 700 acres for himself, as well as looking after another 400 acres for Mr. George Courtauld of Cut Hedge, Halstead. Mr. Walter Whitlock wrote a very prophetic article in the *Agricultural World* dated January 26th, 1901. Here is a brief extract: "I do not see how the owners of smallholdings can go in for steam cultivation and I am convinced that steam cultivation AND MOTOR POWER will be the agriculture of the future."

Mr. Carleton Whitlock goes on to tell me that 73-year-old Mr. Jack Smee still does his stint in the firm's agricultural machinery works. He started in 1900 repairing traction engines and by the year 1916 was at work repairing tractors. One of the firm's earliest tractors was a 1917 Overtime, and in 1918 a Titan arrived at the works in pieces, and was duly assembled. Two years later came their first Fordson, which was supplied to the late Mr. G. A. Goodchild of Great Yeldham. By 1921 the firm were handling crawler tractors and the International chain-driven model. The latter was delivered to the late Mr. Frank Marsh, of Wickham St. Pauls, who was a well-known threshing machinist. Mr. Whitlock says his firm was a little unfortunate with fires, due to tractors. A Titan which was used for driving shafting for the machinery backfired on one occasion and set light to a County Council steam-roller in for repair, the awning of the latter being burnt off. In 1921 a Fordson tractor came in for repairs to the leaking paraffin tank. Jack Smee saw a youngster named Whitbread, and cautioned him to empty out the paraffin before soldering the leaks. Whitbread failed to take this precaution, and as a result the whole of the works were burned down. Whitbread was rushing from one end of the shop to the other, and Jack Smee bowled him over because his back was aflame—merely to get sworn at for his trouble. Mr. Whitlock says for years the firm used an Austin tractor which successfully drove shafting.

Before leaving the subject of fire as applied to tractors, I am reminded of a classic story from Des Moines, in the U.S.A., and which actually happened in 1957. All was set for a lecture on "Farm Safety" when workers and farmers were informed of its cancellation because the lecturer was in hospital with extensive burns. Inquiries as to his welfare revealed an object lesson. He had cleaned down his tractor with petrol, stood back to admire his handiwork and . . . lit . . . a . . . cigarette.

I have tried not to leave out any make which actually was either built in, or brought into, this country up to 1930. The index will show well over a hundred different makes, and through the kindness of Messrs. W. J. Priest and J. H. Milne, of *Farm Implement and Machinery Review*, I am able to fill in one or two gaps. In 1915, for instance, there appeared a small 4½ h.p. petrol tractor called the Service, made in England. She was claimed to develop a drawbar pull of over one thousand lb. There was also an American small tractor called Once-over-Tiller, with an 8 h.p. engine attached to an ordinary riding-plough. It embodied a steel-toothed rotor, set to the right of the share and mouldboard. The rotor was geared to the top of the shaft of a small petrol motor which turned the rotor at about 500 r.p.m. This caught up the newly turned furrow slice and the rotated teeth shred the weeds and soil into a pulverized mass ready for planting.

Three other lesser-known American models which worked here during the first world war were the 1915 Lion, weighing 3,200 lb. with a twin-opposed engine, adjustable governor and on three wheels; the 1917 Kingsway was similar, but weighed thirty two cwt., and could run on either petrol or paraffin, and the third was the Moseley, of which I am able to show a picture. She was not unlike the Fordson, but had an interesting springing to the rear wheels. There were also the Interstate and the Killen-Straite, both or which became popular in Britain. The Hodgson attachment converted a Ford car into a tractor!

The next chapter gives an account of the valuable trials held at Aisthorpe in 1920.

VII

THE AISTHORPE TRIALS OF 1920

THE Royal Agricultural Society has always encouraged the inventive genius of the agricultural engineer. It is therefore not surprising to note that the very next year after the South Carlton trials of 1919 found them organizing an important tractor demonstration at Aisthorpe and Scampton, nr. Lincoln. The aerodrome at Scampton provided excellent facilities for storing the entrants and for the necessary administrative offices. About eight hundred acres were involved, the fields were measured up and four similar plots were allotted to the several machines of each class. The trials were competitive and classes were as under:

Class 1. Internal-combustion direct traction not exceeding 24 h.p.

Class 2. Similar, but not exceeding 30 h.p. and to take three furrows.

Class 3. Over 30 h.p. and taking four furrows.

Class 4. Direct traction steam engine.

Class 5. Double engine cable tackle internal combustion.

Class 6. Double engine cable tackle steam.

Class 7. Self-propelled plough.

There were quite stringent rules and conditions, and each class worked on both light and heavy land. Hilly land was also brought into the contest and tests of stopping uphill and downhill were also carried out. The overall ability to turn in prescribed circles was measured, and the tractors with power pulleys were tested on the actual driving of machinery. The machines were critically examined and were even weighed before the final judging.

It might be of interest to record the judges' names. They were W. E. Dalby, Harry Buddicom, W. R. Fieldsend, F. W. Lanchester, R. M. Greaves, B. Howkins, L. A. Legros, H. Riall

Sankey, Bayntun Hippisley, Henry Overman, Dunbar Kelly, and Fred Scorer.

As in the 1919 trials, the National Physical Laboratory dynamometer was in use for recording the drawbar pulling power of the tractors under test.

A most interesting feature of these trials was the inclusion of classes for steam engines, but in the class for direct traction, Mann's steam wagon, mentioned in Chapter III, was the only entrant. Fowlers also exhibited their well-known steam cable ploughing, as described in *Traction Engines*. Both Fowlers and McLarens had also motor cable sets at work and these were of considerable interest. The introduction of lightweight high-speed oil engines as prime movers revolutionized the cable cultivation system; there was the flexibility of steam at about one third the weight. During the first world war Messrs. Walsh & Clark, of Guiseley, Bradford, had, in fact, marketed the Victoria oil ploughing engines for cable work. In appearance these were very like steam engines, even having chimneys for the exhaust fumes. It was stressed that these engines did not travel on the land, as they were working, but merely on the headlands. This was just as well, for each engine weighed five and a half tons and had twin cylinders running on paraffin.

Fowler's model was termed the "motor cable plough". The engines were four-cylinder and had electrical self-starting complete with dynamo. The fuel in their case was petrol and their h.p. rating was 46. The ploughing gear gave a choice of two speeds on the cable, and, as in the steam plough, the drums revolved horizontally. Before this time Fowlers had produced a "patent motor ploughing engine" which in appearance and design very closely resembled their steam engine—even to a chimney and "dummy" smoke-box. My illustration shows how very close was this resemblance.

The other famous Leeds firm of J. & H. McLaren called their machine McLaren's Motor Windlass, and it was a wonderful job. Mr. Phil Scott, of Morley, tells me that some of them are still in use in Yorkshire, and large numbers were exported, of course. Each windlass consisted of a very strong frame of steel channels rigidly stayed and carried on four steel road wheels. Chain steerage was attached to the front wheels and worked from the footplate and the hind wheels were actuated by a worm drive through a differential gear. These engines could develop 60 to 70 h.p. There was an auxiliary starting engine and gear, comprising a twin water-cooled petrol engine of 8 h.p. to facilitate starting the big diesel engine from cold. Compared with the steam ploughing engine the absence of a steam boiler enabled the makers to locate the winding drum in a vertical position. The great advantage here, of course, was added strength, for the drum was carried on a shaft held at both ends and not on a stud as in the horizontal steam-driven drum. The coiling of the rope was also simplified, as it was not affected by the slope of the ground. This machine could, in fact, stand and work at any angle, and the rope draught was low to prevent side-slipping. The chain drive was employed instead of a gearbox, and a variety of drum speeds were offered, enabling the most difficult soils and overseas conditions to be tackled. The general appearance of these engines was quite modern, as will be seen from the photograph.

As far back as 1893 McLarens had experimented with a rotary tiller, and although other firms had tried after that date, no really satisfactory machine had been evolved. McLarens stated quite frankly in their catalogues that efficient cultivation demanded the plough and could not be obtained with a "once over" system of rotary tillage.

In the year 1927, however, the intrepid Fowler firm produced their 225 h.p. Gyro-tiller, which took 10 ft. of land at a time and by rotary system pulverized the land up to a depth of 22 in. These machines were built to Storey's patents and two rings of ploughs rotated on a vertical axis. Carried on a massive hinged frame, the unit could be lowered or raised. The engine was a six-cylinder, and some of these fascinating machines are still used by Bomford Bros. Ltd., Evesham. Mr. Alf Pepper, of Fowlers, who sent the two photographs, tells me the first model was built in 1927 and weighed over twenty-three tons. The engine was a Ricardo petrol engine, but because it needed fourteen gallons of petrol per hour an M.A.N. 137 h.p. diesel engine subsequently replaced it. Once over the ground, the Gyrotiller pulverized and aerated the soil to a depth of 18 in. and a two- or three-row ridger could be fitted behind. The land could then be planted, and such a method proved invaluable for the sugar-planters of Cuba and elsewhere. Germination of the cane plant was two or three weeks earlier by the use of the Gyro-tillers. Subsequently, by 1930, Fowlers were making an 80 h.p. Gyro-tiller as well as a "Tractiller" on a similar principle save that this was a separate unit hauled by a tractor.

The Austin Motor Company had two tractors on trial and, as they were similar to the model described at the 1919 trials, it is pointless to repeat the specifications.

Fabrique d'Automobiles Berna, of Switzerland, sent their Berne tractor. This was a four-cylinder petrol-fuelled traditional type with a seat resembling that of a car. There was a differential gear, and exchangeable rubbered wheels were offered for road work.

Blackstones sent a similar model to that already described, but the Boon was a newcomer. This tractor was built by Ransomes of Ipswich and was a twin-cylinder paraffin-driven machine weighing fifty-two cwt. It drew a three-furrow Ransomes plough, yet I have talked to scores of East Anglians who do not recall this machine.

Ruston & Hornsby Ltd., of Lincoln, had two models of their British Wallis present. It was a four-cylinder 23 h.p. paraffin tractor with two forward speeds, and weighing thirty-eight cwt.

From Wisconsin, U.S.A., came two Case tractors similar to the one already noted, and from Canada the Chase Tractors Corporation sent a Chase tractor powered by a four-cylinder Buda motor with vertical cylinders cast in one block. Running on paraffin, this outfit weighed forty-nine cwt.

Another U.S.A. entry from the Cleveland Tractor Company, of Ohio, were two Cletrac chain-track machines. As is seen by the photograph, these were almost modern in appearance. Rated at 22 h.p. with a four-cylinder engine, they each weighed thirty-three cwt. The Essex-built versatile Crawley Agrimotor was there, of course, but has already been described, as has the E.B. (Emerson-Brantingham) from the U.S.A., and the Italian Fiat.

The famous Fordson also came along and was then being manufactured in Cork, Ireland. We have described the Fowler motor cable and steam cable entries, but this firm also had a Fowler 20 h.p. motor plough, very like the Crawley in appearance.

Their own plough and cultivator were used with this machine which weighed forty-one cwt.

Both the Garner and G.O. (General Ordnance) were there, as well as the Glasgow, but a newcomer was the Hart-Parr, built by the Hart-Parr Company, of Charles City, Iowa. The twin-cylinder engine developed 28 h.p. running on paraffin, the total weight being forty-nine cwt.

Another American newcomer was the John Lauson from New Holstein, Wisconsin, a four-cylinder tractor running on paraffin, weighing sixty-two cwt. Mr. Norman R. Collings, of Collings Bros., of Abbotsley, Huntingdon, tells me he drove this tractor at these trials and it will be seen he was in the prize list. He has had a lifelong experience in the tractor world.

We have mentioned McLaren's cable outfit as well as the Mann steam tractor (sole entrant in its class). As in the other trials, there were two Martin entries and a Moline. Weeks's New Simplex also worked at Aisthorpe, but the Parrett from the U.S.A. was a newcomer. She presented an unusual appearance, with larger front wheels than normal. With a four-cylinder paraffin engine and three forward speeds, she weighed forty-six cwt.

A new British tractor, the Peterboro', was built in Peterborough by Peter Brotherhood Ltd. A four-cylinder paraffin tractor weighing fifty cwt., her engine was completely enclosed. The modern-looking Pick was there from Stamford, but this, too, has already been described.

From Janesville, U.S.A., came the Samson, a four-cylinder tractor with belt pulley and weighing only thirty cwt. The firm (Samson Tractor Company) also made a three-wheeled Sieve Grip, whose special feature was cast-steel sieve-type wheels, which were designed to minimize soil-compression. I am able to give a photograph of this latter design. The other interesting feature in the Sieve Grip was provision for the cleaning of all the air indrawn by the engine. Dust, chaff, and impurities passed in through a water bath. The three-wheeled tractor weighed forty-eight cwt. The model M at the trial, however, had four wheels and weighed only thirty cwt.

The Santler was there again and, of course, Saundersons had three of their Universal tractors working.

Pick three-wheeler made at Stamford.

The little Garner on trial at Lincoln.

1929 Massey-Harris four-wheel drive.

Early Case followed by Cletrac.

International Junior 1919.

The 1915 Overtime.

1917. The first Fordson arrives in England.

1929 Model Fordson.

From Minneapolis came a new American tractor with a peculiar name—Twin City. It was very like the Fordson in appearance; with a four-cylinder 28 h.p. engine, she had overhead valves, ran on paraffin, and weighed about forty-three and a half cwt.

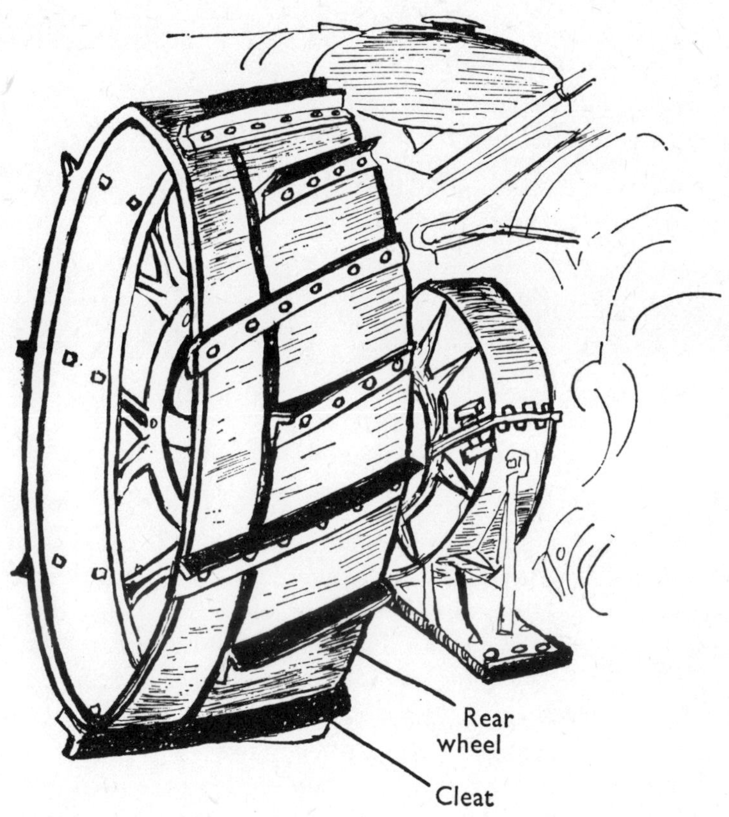

Fig. 16. Samson extension band

In the trials the Ruston & Hornsby double-furrow plough—a one-man outfit—was used. Control was from the tractor-driver's seat by means of a cord, and it was convertible to three furrows. The other plough was Ransome's three-furrow plough, again a one-man, self-lift, cord-operated type.

After gruelling tests, the judges' awards were:

Class 1. (1) Gold Medal and £20. J. I. Case Company.
(2) Bronze Medal and £10. H. G. Burford (Cletrac).

Class 2. (1) Gold Medal and £20. Ancona Company (British Wallis).
(2) Bronze Medal and £10. Peter Brotherhood Ltd. (Peterboro').

Class 3. (1) Gold Medal and £20. John Lauson.
No second prize awarded.

Class 4. (1) Gold Medal and £20. Manns' Steam Cart. (Only entry.)

Class 5. (1) Gold Medal and £20. John Fowler & Co., Leeds.
(2) Bronze Medal and £10. J. & H. McLaren, Leeds.

Class 6. (1) Gold Medal and £20. John Fowler & Co., Leeds.

Class 7. (1) Gold Medal and £20. Crawley Agrimotor Co., Saffron Walden.
(2) Bronze Medal and £10. Motrac Engineering Co. (Moline.)

The judges' report was very full, and a good deal of it seems rather irrelevant today. At the same time, it helps appreciation of the work of these pioneers if we consider a few points the judges made. They pointed out, for instance, that farm tractors had to be versatile and do a great many jobs other than ploughing, cultivating, and harvesting. By this time the more adventurous farmers were drilling their corn by tractor power and doing a lot of haulage and stationary belt work. Furthermore, these machines were being used in remote parts of the country, far from well-equipped workshops, driven by unskilled labour and driven in all weathers. The judges thought it creditable, therefore, that forty-six tractors had been entered and thirty-eight had actually worked in the trials, only two of the latter having to retire. The uniformity of excellence had indeed made judging very difficult. The trials also took into consideration the cost of ploughing, fuel then being very much cheaper than it is today.

Of the thirty-six competitors which completed the trials, four were supported on chain-tracks, all the American and Canadian machines used paraffin, four tractors of either British or Continental manufacture used petrol alone, and two were, of course, steam propelled on coal.

The preliminary use of petrol for starting was usual in these paraffin engines and many had a special heating device formed as a jacket on the exhaust pipe. Many of the U.S.A. tractors had an impulse-starter fitted to the magneto ignition system. Two tractors were indeed fitted with self-starters, which is remarkable in that many cars in use had not that luxury in 1920, as I knew from personal experience.

The Ricardo engine which powered the Peterboro' tractor had a clever device which turned a small portion of the exhaust gas direct into the carburettor, thereby warming the intake air. In the Moline engine, a small fraction of the fuel oil contained in the incoming air was allowed to trickle unvaporized into a pocket in the carburettor. In this pocket was a sparking plug and a supplementary supply of air was introduced into the pocket. This meant that there was continual combustion in the pocket of this small fraction of the charge; the heat produced was used to raise the temperature of the charge and to vaporize the fuel.

Many of the tractors were fitted with either dry separators or water-washing arrangements for cleaning the intake air. These cleaners prevented stoppage or trouble in jobs like harvesting, where dust and chaff were so prone to be drawn into the induction pipe. Some of the tractors were fitted with governors, and sometimes their action had to be supplemented by hand throttles, especially when turning at the headlands or when manoeuvring. This has, of course, been the cause of many accidents, even in these modern 1960s.

Many models had indicating devices to show that the oil level in the crankcase had not fallen below a safe working level. The weather was very wet apparently during the latter part of the trials, which made the successful work done a subject of very great credit to all concerned. The R.A.S.E. gave great credit to the Society of Motor Manufacturers and Traders for help in stewarding and making many of the arrangements.

The most suitable conclusion to this chapter seems to be the enumeration of all other known makes of tractor either built in Britain or imported, up to 1920. Naturally, I will only deal with those which have not been mentioned in the former chapters.

There was, for instance, the little Beeman, a forerunner of many small garden and market-garden tractors. She was a London-made miniature single-cylinder model of only 1¼ h.p. and, of course, had limitations.

The British American Import Company, of Haymarket, were distributing the Eagle at this stage. She was a 24 h.p. twin-engined tractor weighing forty-five cwt.

The Farmer's Boy was an American lightweight (sixteen cwt.) powered plough, behind which the user walked and used the handles for control as in a horse plough. She was a four-cylinder three-wheeled machine of 12 h.p., sold by Morris Russell of Great Portland Street, London.

The Heider was made in Rock Island, U.S.A., and sold in England by Willys-Overland. The Kingsway was another lightweight three-wheeled tractor made in two sizes: (*a*) for two-furrow work, priced at £240, and (*b*) for three-furrow ploughing, costing £350. This tractor was sold by Gaston, Williams & Wigmore, of Great Portland Street.

The Lightfoot was a small four-cylinder track-layer of 6 h.p., weighing only twenty-five cwt. It was primarily for the market-gardener, and was built in Coventry. A four-cylinder four-wheel tractor of forty-six and a half cwt. was the Little Giant, which came from Marlboro' Motors Ltd., of Euston Road, N.W.1, as did the Lion. The famous traction engine firm of Marshall, Sons & Co. Ltd., of Gainsborough, now had a Marshall tractor on sale. She was a big machine, very like a traction engine in appearance, and was run on diesel oil. By complete contrast was the little Morse, a four-cylinder, four-wheel tractor of 32 cwt., made by Fairbanks, Morse & Co., of Southwark Street, London. Power Farm Supply Co., of Coventry, were importing from Chicago the Monarch-Neverslip. She was built in three sizes and was on tracks. Weighing five tons, with a horsepower rating of 12–20, she would be very heavy for British clay soils, but could run on either petrol or paraffin.

Richard Garrett & Sons Ltd., of Leiston, Suffolk, had their steam tractor Suffolk Punch, which is described in *Traction Engines* and was very similar to the Mann steam cart.

Rejax Ltd., of Tottenham Court Road, were marketing a unit for converting Ford model T cars into tractors. The Rejax cost £50 and was like Eros, which I have already noted.

The Worthington Company, of Kingsway, W.C.2, brought out their Worthington-Ingeco in this period.

VIII

DEVELOPMENT UP TO 1930

In May 1927, three years after I had qualified at Chadacre Agricultural Institute, I went back to one of their fields as one of the junior members of the staff of Mann Egertons, Bury St. Edmunds, tractor division. The object of the visit was a mole-draining demonstration with tractor power, arranged by the West Suffolk Agricultural Committee. One of the heaviest fields had been chosen, stiff clay with chalk and plenty of flint. Until this time neither I nor most of the others present had ever seen mole draining tackled except by steam power. The mole ploughs tested were a Darby, a Ransome, and a Cavanagh Excavator. The Darby was tried at a depth of fourteen inches, but one Fordson could not move it. Two Fordsons were then tried out, first abreast and then tandem, but they kept on lifting on the front wheels and could not be said to do the job satisfactorily. The two tractors were then hitched to the Ransomes mole plough at a depth of twenty and then sixteen inches, but could not pull it. The late Spencer Debenham, of Lawshall, dashed home and fetched his Burrell traction engine and pulled the mole plough with success at twenty inches.

The Cavanagh outfit was different and consisted of an excavator and mole plough. Drawn by a Fordson tractor the excavator first made a narrow trench twelve inches deep by taking out two or three inches at a time. After this, the mole was drawn through a further eight inches deep. The big flints encountered proved very detrimental to the satisfactory working of this implement which, it was claimed, could also draw trenches for pipe draining. The whole experiment was, however, premature and proved that mole draining heavy land with tractor power was not practical. It is only fair to record that with track-laying tractors this job has for some years been done successfully since that time.

The World Tractor Trials of 1930 were of great importance,

and as they cover the limit of the scope of this book, I hope I shall be forgiven for mentioning them in detail. Sponsored by the Royal Agricultural Society of England in conjunction with the Institute of Agricultural Engineering of Oxford University, the trials attracted some good machines. The ultimate object of the trials was similar to that of other trials, i.e. to display the latest developments to any interested parties. But an attempt was made to increase the value of such a display by first subjecting each entrant to severe tests.

Thirty-three tractors and three market-garden cultivators were entered, twelve coming from the U.S.A., eight from Britain, five from France, four from Germany, two from Sweden, and one each from Hungary and Ireland. The market-garden cultivators were tested at Evesham in May, and the other tractors at Wallingford from June 2nd to July 26th, each one doing at least twenty-four hours work. Only three machines failed to stay the course, two on account of mechanical breakdown and one because of changed plans on the part of the makers.

The tests were not made competitive, as the organizers preferred to try to show exactly what each tractor could do under normal working conditions. Ransomes, Sims & Jefferies Ltd., of Ipswich, and James & Frederick Howard Ltd., of Bedford, lent implements for the field work. Here is the list of entrants with a few details:

Austin. A wheel tractor with four-cylinder engine, nominal h.p. 12/20, starting and running on petrol. Entered by Société Anonyme Austin, of Rue de Lafayette, Paris. Demountable rims with spade lugs, three-speed gearbox, weighing 4,060 lb., and priced at £210. The firm had a similar model on trial, but running on paraffin.

Aveling & Porter. An interesting tractor made by a famous traction-engine firm. Wheel-type with four-cylinder diesel engine, 22/38 h.p., starting by electric motor. She was entered by Agricultural & General Engineers Ltd., of Aldwych, and had spade lugs, road rims in one section, a three-speed gearbox, plate clutch, and weighed 7,300 lb. Price £525.

Blackstone. A wheel-type tractor with a four-cylinder 20 h.p.

diesel engine, started by an auxiliary petrol engine. This had a three-speed gearbox and multi-plate dry clutch, weight 7,300 lb., with spade lugs, road rims in two halves. Price £500.

Case (Model C). A wheel tractor with four-cylinder paraffin engine (started on petrol), built in U.S.A., 17/27 h.p. Here was a three-speed gearbox with twin-disc clutch. Spade lugs with scrapers and road rims in two halves, weight 3,700 lb. Price £248.

Case (Model L). A wheel tractor with four-cylinder engine running on paraffin (started on petrol), 26/40 h.p., weight 5,200 lb. She had a three-speed gearbox and twin-disc clutch, with similar wheels to the other model, and was priced at £348.

Caterpillar Ten. A track-layer with four cylinders, starting and running on petrol, 10 h.p., manufactured by the famous Caterpillar Company of California. Three-speed gearbox, brakes on each track, and weighing 4,420 lb., selling at £290.

Caterpillar Fifteen. Similar details as above, but 15 h.p., and weighing 6,020 lb. Price £390.

Caterpillar Twenty. Similar details, but of 20 h.p., weighing 7,670 lb., and selling at £505.

Caterpillar Thirty. As above, but weighing 10,160 lb., and priced at £645, 25/30 h.p.

The final specimen of this remarkable range was the

Caterpillar Sixty. Developing 50/60 h.p., weighing 19,450 lb., and priced at £1,150, a contractor's model rather than a farmer's unit.

Citroën-Kegresse. An entry from France. The firm also sent a duplicate entry. She was a track-laying tractor with four cylinders, 13 h.p., starting and running on petrol. Here was a six-speed gearbox and the tractor weighed 3,810 lb. Cost £462.

Fordson. A wheel-type four-cylinder engine of 11/20 h.p., starting on petrol, running on paraffin, built in Ireland and marketed from Trafford Park, Manchester, already described elsewhere, but strangely enough the entry at these trials suffered a major breakdown.

Hofherr-Schrantz. An all-steel crude-oil tractor, made in Budapest, Hungary; single-cylinder diesel engine with hot-bulb

Fordson replaces two men and six horses.

Ancient and modern.

"Eros" attachment converting Ford touring car into tractor.

The latest Fordson Dexta.

ignition, 20/30 h.p., started by blowlamp, weighing 6,600 lb., two-speed gearbox, and priced at £350. This firm was in association with Clayton & Shuttleworth Ltd., of traction-engine fame, who looked after their interests in this country.

International Farmall. A wheeled tractor with a four-cylinder engine rated at 9 h.p., starting on petrol and running on paraffin; spade lugs, one-piece road rims, three-speed gearbox and single-plate clutch; weight 4,120 lb., and selling at £220. The makers, International Harvester Company, had two other models present.

International 10/20. A very similar tractor in many respects to the Farmall, the engine being enclosed to a greater degree. The price was the same, but it was a slightly heavier machine.

International 15/30. A more powerful example of the other two types, weighing 6,720 lb., and priced at £320.

Latil. A four-cylinder tractor made in France, and with a four-wheel drive; rated at 17 h.p. and running on petrol, she had a six-speed gearbox on pneumatics, weighed 5,970 lb., and cost £655.

Lanz. A wheeled tractor with a single-cylinder semi-diesel engine, starting by blowlamp; made in Germany and first marketed in this country by J. Mann & Son, of Saxham, Suffolk. Two models appeared at the trials; they were rated at 15 h.p., had a three-speed gearbox with a three-jaw clutch. Weighing 6,200 lb. and priced at £325, this was in every way a forward-looking tractor, with a very apt name, the Bulldog.

Linke. A track-laying tractor with a four-cylinder engine, rating 35/45 h.p., starting and running on petrol. She was a modern-looking tractor made in Germany, with a three-speed gearbox and single-disc clutch, priced at £530, weighing 7,620 lb.

Marshall. A wheeled tractor with a single-cylinder diesel engine, started by torch and running on diesel oil. She was 16 h.p., with four forward speeds, and was a forerunner of a very good line of tractors built by the famous traction-engine firm of Marshall, Sons & Co. Ltd., of Gainsborough. Price £315.

Massey-Harris. This firm had two models, a 20 h.p. and a 30 h.p. Both were wheeled tractors, running on paraffin, with four cylinders; the wheels had spade lugs and road rims in two sections. The smaller tractor had a three-speed gearbox and its bigger brother a two-speed ratio. The smaller model sold at £220 and the big one at £300, both being made in the U.S.A., respective weights being 3,590 lb. and 4,690 lb.

Mercedes-Benz. This was another single-cylinder wheeled tractor, starting by torch and running on diesel oil. The wheels had strakes and the road rims were in two sections; three-speed gearbox and single-drive clutch. Although built in Germany, the tractor was sold in this country by J. & H. McLaren Ltd., of Leeds, price £360; weight 6,100 lb.

McLaren. This was a twin diesel wheeled tractor with torch starting, three-speed gearbox, and a single dry-plate clutch. She sold at £500, was built by the Leeds traction-engine firm, and weighed 6,090 lb.

Munktells. This tractor was a twin diesel type, but was started by built-in blowlamps. There were two models of this Swedish-built tractor present, one 15 h.p. and the other 20 h.p.; wheels in each case had spade lugs and road rims. The lower-powered model weighed 6,010 lb. and sold at £270, and the larger one, weighing 6,830 lb., was priced at £340.

Peterboro'. Made by Peter Brotherhood Ltd. A British machine with four cylinders, running on paraffin and rated at 22 h.p.; quite a good tractor, but became a "casualty" at these trials.

Rushton. The Rushton Tractor Company, of Walthamstow, E.17, had two interesting tractors. One was termed the Roadless and, as its name implied, was a trackless type of 18 h.p. This tractor had four cylinders and ran on petrol, with a three-speed gearbox. Weighing 5,650 lb., she sold at £434.

The other model was a wheeled tractor of 14 h.p., with four cylinders, and the wheels had cleats with scrapers plus road rims in sections. Weighing 3,950 lb. she cost £209. 10*s*.

Vickers. This famous firm built an Aussi tractor, with four cylinders. Rated at 23 h.p., she started on petrol and ran on paraffin. She was built at Crayford, Kent, but did not perform at these trials.

There were three motor cultivators and they must be well worth mentioning because of the great popularity of such types of rotary hoes and garden machines today. All three were entered by Messrs. George Monro Ltd., of Waltham Cross:

Monotrac. This was a 3 h.p. single-wheel petrol-driven cultivator; the operator walked behind and steered by the handles as in a motor mower. At £50 this was good value; it weighed 240 lb.

Duotrac. This was a $2\frac{1}{4}$ h.p. model on two wheels and with a belt pulley at the front, priced at £38.

Rototiller. A 4 h.p. fuel-powered machine on the now popular rotary principle, selling at £98 and weighing $3\frac{1}{2}$ cwt.

Three tractors unfortunately were withdrawn from the trials: the Vickers because of the firm's change of programme; the Fordson, strangely enough because of a cracked cylinder block attributed to a fault in the casting, and the Peterboro' because of damage to piston, connecting rod and splash guard. This was thought to be due to a nut becoming loose in the crankcase.

There were three broad tests laid down for each tractor, belt tests, drawbar tests, and field tests of ploughing and cultivating. Where suitable, there was an additional test of road haulage. To carry out the belt testing, an electro-Dynamometer was used. It was an electrical generator with a separately excited field and a controllable bank of resistance. The fuel consumption of each tractor during its belt test was also carefully measured.

In the case of drawbar testing something similar was put into use in the form of "dynamometer cars". They consisted of a three-wheeled chassis, the front for steering and the rear wheels typical tractor wheels. The latter, in fact, corresponded to the driving-wheels of a tractor, and as these revolved they drove an electrical generator through a four-speed gearbox and final chain drive. As in the case of the belt dynamometer a controllable resistance was provided for the purpose of putting any desired load on to the generator. By means of switches the car was made either harder or easier to pull. The four-speed gearbox allowed the generator revolutions to be kept at a similar value over a wide speed range along the ground.

To ascertain the power exerted by the tractor it was necessary to measure the drawbar pull of the tractor and the speed at which it worked. This was done by means of the "traction dynamometer", in two parts—the link or pressure unit, and the recording unit. The former was merely an oil-filled cylinder fitted with a close-fitting plunger. The plunger was hitched to the tractor and the cylinder connected to the car. The oil pressure thus set up was transferred through steel piping to the recording unit, where it caused a much smaller plunger to move. The second plunger, controlled by a spring, carried a pen which marked an ink line on a moving roll of paper, driven at constant speed by a clockwork motor, the displacement of the pen being controlled proportionally by the pull of the tractor.

The ploughing tests were quite rigorous and carried out on grassland, where there was chalk with clay beneath. All ploughing was done at 4½ in. depth, and the same land was afterwards used for cultivating tests. Each tractor did four hours of each operation; no attempt was made to load the larger tractors to their fullest capacity and no measurement of power was taken. Each tractor apparently performed quite well and only two stoppages for minor repairs are on record. All the results of the belt and drawbar testing were finally tabulated, the tractors being divided into groups depending upon what fuel they used. Even the cost of the work was computed in relation to fuel and oil consumption. Any farmer who wanted to work using one of the tested tractors was advised to decide:

1. How many hours each of belt and drawbar work per annum could he be expected to carry out.
2. What percentage annual allowance on the first cost should be made for interest, depreciation, and repairs.
3. What proportion of the driver's annual wage should be charged against the tractor's performance.

By then making reference to the final tables, a prospective buyer would be able to choose his tractor.

Only two tractors entered the road-haulage test, the Latil and the Citroën. In each case they had to draw a loaded trailer over 30 miles of public road. The Latil took ten tons 28 miles in 1 hour

52 minutes at an average speed of 15.3 miles per hour. The route involved a gradient of one in thirteen.

The Citroën-Kegresse hauled four tons for the 28 miles in 2 hours 25 minutes at an average speed of 11.8 m.p.h. with a similar gradient *en route*. Both these tractors performed well.

The little market-garden cultivators worked during May 1930 on a market-garden estate at Pitchill, nr. Evesham, and each did forty hours' running.

The judges found that the Monotrac was suited for hoeing between rows and produced its best work on even, well-worked soil, such as is usual in gardens and nurseries. It was not so easy to manipulate where the soil was uneven. The engine was powerful enough for the job.

The Duotrac was easy to manipulate and was said to be suitable for any cultivation between rows. A ground clearance of ten inches allowed the machine to work astride of rows in some places.

The Rototiller did tilling, rolling, ridging, hoeing, and deep cultivation, and apparently made a very good job of it, too.

The year 1930 appears to be the period when this survey should cease. The subject, after all, is "old" farm tractors, and in the concluding chapter I want to mention briefly the way in which the trends became "modern", as well as mentioning a few of the veterans which are still maintained by their owners in working condition.

IX

SOME VETERANS STILL WORKING AND PRESENT-DAY TRENDS

In this final chapter I want to give a few details of enthusiasts who are still using some ancient tractors and of others who are preserving them. Mr. Dennis Fogerty, of Fordingbridge, tells me that during the second world war he worked for a Mr. Hayward at Coryton, Devon. They were doing itinerant threshing with two old Saunderson tractors which had been slightly modified. A special gearbox had been fitted and the rear wheels were on pneumatic tyres. Every half-hour the radiators had to be topped up, as they were always boiling. Although running on T.V.O. as fuel whenever a steep hill was approached it was the custom to turn over to petrol for a better pull. Mr. A. C. Wright, of Raskelf, Yorkshire, tells me he himself has memories of early tractor work in the first world war, and says a Titan was driving a saw bench in his village until very recently. Quite near to me in Essex is a Massey-Harris 1926 tractor, owned by Mr. H. Metson, of High Ongar, and in continuous use, and at Bradfield Combust, Suffolk, Mr. Bates still uses an old International Junior. A veteran of the same make recently retired at Ficle, nr. Lewes, Sussex. Mr. Cyril Phillips tells me it was a 1913 model which his father bought for £11. At Gloucester, Mr. C. Nichols tells me his son has a 1914 McCormick and a 1918 model of the same make which he is reconditioning. John Hirons, of Witney, has a 1916 Titan and a 1920 International Junior in good trim. Mr. T. A. Hinch, of Little Casterton, Lincolnshire, has a big collection of veteran tractors which he overhauls in a very modern workshop. He has the following veterans: a 1917 Titan, restored and painted; a 1920 Fordson; a 1917 Clayton crawler in like condition. In addition he has a 1921 Austin in going order, a 1926 Caterpillar being restored, and a 1917 Saunderson, usable but unpainted. He modestly tells me that the others are "in the rough", and include a 1920

International Junior and a 1910 Mogul single cylinder. The Clayton crawler is very unique, of course.

In Crawley, Sussex, Mr. J. Elles has a very old Massey-Harris tractor still working.

At Kirton Holme, nr. Boston, Lincolnshire, Mr. F. A. Smith has a good collection of veterans—fifteen makes are included, namely, Mogul, Titan, International Junior, Overtime, Saunderson, Fiat, Rushton, and Lanz, some of these being in duplicate. Mr. Derek Hackett, of Bridstow, nr. Ross-on-Wye, has a quite remarkable collection and they are: Avery 1916, Austin 1919, Crawley 1920, Fiat 1919, Ivel 1903, International Junior 1919, Massey-Harris four-wheel drive 1929, Mogul 8/16 of 1915 vintage, Mogul twin 1913, Mogul single cylinder 1917, Moline 1917, Parrett 1917, Overtime 1919, Rushton 1929, Saunderson 1917, Saunderson 1920, Titan 1916, Wallis Cub 1917, and a British Wallis 1920.

Some years ago, the late Mr. D. N. McHardy, of *Practical Power Farming*, made a collection which is now housed at Shuttleworth College, Biggleswade. These tractors were exhibited at the Bath and West Show at Dorchester in 1951. They are as follows: Austin 1923, Cleveland Cletrac 1918, Fo dson 1917, Overtime 1915, Peterboro' 1919, Saunderson 1914, and Titan 1917.

Mr. K. Stenner of Eggesford, Devon, has a 1915 Titan still in use. So also has Mr. Dick Joice of Anglia T.V. fame, whilst Mr. Don Gordon of Sutherland, Scotland, has a pre-1914 Saunderson.

At the 1961 Royal Counties Show, Mr. E. G. Goldup, of Fernhill, Surrey, exhibited a 1917 Overtime which he has reconditioned from scrap for the owner Mr. J. Perkins of Earlswood (see frontispiece). At Hockwold, nr. Thetford, Mr. F. H. Allsop has one of the earliest Overtime tractors, in regular use driving a mill, and reckons to grind over a thousand sacks a year with her.

The late Mr. Ernest Edney, of Horndean, Hampshire, was apprenticed at Messrs. Saundersons in the first world war; he left in 1917 and brought with him a Saunderson for farm work which was in constant use day and night. Subsequently the lighter Fordsons replaced the Saundersons, says Mr. Edney, and many other farmers did a similar change-over at that time.

It will have been seen that the early tractors were little more than mechanical horses, whereas today's tendency is to fit the tractor to the implement in such a way as to form a single unit. The work of the late Harry Ferguson should be noted in this respect. Born on a farm in County Down, Northern Ireland, Ferguson, at the age of sixteen, had set up a small works in Belfast, where he sold and serviced cars and motor cycles. After trying his hand at car and motor-cycle racing, he turned his attention to aeronautics, and in 1909 he designed and flew a small monoplane, the first "heavier-than-air" machine to be flown in Ireland. He returned to cars in 1911 and then moved towards tractor pioneering. During the first world war he was appointed to maintain all farm tractors in N. Ireland, to guard against a breakdown in food production. He saw the cumbersome implements of that age as a challenge to his inventive ability. Tractors he saw as heavy and difficult machines to handle, but probably the greatest fault to him was the fact that implements were trailed instead of being mounted as an integral part of the tractor.

In 1920 Ferguson designed his first three-point linkage for attaching implements to the tractor, which meant that the driver no longer had to leave his tractor to adjust the implement.

From this beginning eventually came the famous Ferguson system which had a profound effect upon farming and the tractor industry. By 1935 the system was perfected, a combined linkage and hydraulic control system, applicable to a wide variety of farm implements. This development formed a basis pattern which has been followed by tractor manufacturers the world over. Within a year the Ferguson tractor was being built in a Huddersfield factory; prototypes had already been in hand in 1933 at Belfast. Production at Huddersfield ceased with the approach of the second world war, and after Ferguson's famous handshake agreement with Henry Ford, the tractors were made in America. In 1945 Ferguson arranged for his tractors to be built by the Standard Motor Company at Coventry. At a cost of £5,000,000 the factory was tooled up, and after some 306,221 tractors had been produced, there came a break in relations between Ford and Ferguson. This led to the famous lawsuit when Harry Ferguson alleged that Henry Ford the second (grandson of his former partner), the Ford Motor

Company, and others, had "deliberately copied the Ferguson system tractor and farm implements, unlawfully seized and used Ferguson's inventions, engineering developments, designs, and ideas, violating the agreement of trust and confidence made in 1939 by Harry Ferguson and Henry Ford".

After four years, Harry Ferguson Inc. was awarded £3,303,570, believed to be the largest sum ever awarded by final judgement in a patent action. The award was for past royalties on Ferguson tractor patents from July 1st, 1947, and the Ford Company were ordered to cease making the tractor, as then constructed, by the end of 1952. Ferguson always maintained that it was not for money that he maintained this long and strenuous fight. He declared it was his intention to protect the rights and interests of young and go-ahead companies which develop new inventions from being overridden by powerful and established organizations.

In August 1953 Mr. Ferguson announced the amalgamation of his companies with Massey-Harris Ltd., of Toronto; it was a natural amalgamation because of the complementary rather than the competitive nature of the two companies' activities. Massey-Harris, with 105 years of background, fourteen great manufacturing plants, 19,000 employees, and a dominant position in the world markets, was a full-line company. Over the years it had pioneered the famous self-propelled combine-harvester which was revolutionary in all parts of the world where wheat was grown.

It is beyond the scope of this book to trace modern tractor development except perhaps to note the tendencies which continue to evolve. The very small light tractor hoes (especially rotary hoes) have been increasingly improved and are, of course, invaluable for gardeners, nurserymen, and market-gardeners. The same can be said for the wheelbarrow type of tractor with its variety of implements. In the bigger field there has been an increased attention to drilling with accuracy and with correct spacing, and the application of power to row-crop work has increased enormously. Reversible ploughs are now common. Crawler tractors have always been of value, as they can often operate under conditions which defeat a wheeled tractor. These now vary from the small 6 h.p. crawler to a giant capable of ploughing with eight furrows on heavy land.

INDEX